Denis Stein

Mobile laser scanning based determination of railway network topology and branching direction on turnouts

Schriftenreihe
Institut für Mess- und Regelungstechnik,
Karlsruher Institut für Technologie (KIT)

Band 038

Eine Übersicht aller bisher in dieser Schriftenreihe erschienenen Bände finden Sie am Ende des Buchs.

Mobile laser scanning based determination of railway network topology and branching direction on turnouts

by
Denis Stein

KIT Scientific Publishing

Dissertation, Karlsruher Institut für Technologie
KIT-Fakultät für Maschinenbau

Tag der mündlichen Prüfung: 24. Oktober 2017
Referenten: Prof. Dr.-Ing. Christoph Stiller, Prof. Dr.-Ing. Dirk Abel

Impressum

KIT Scientific Publishing

Karlsruher Institut für Technologie (KIT)
KIT Scientific Publishing
Straße am Forum 2
D-76131 Karlsruhe

KIT Scientific Publishing is a registered trademark
of Karlsruhe Institute of Technology.
Reprint using the book cover is not allowed.

www.ksp.kit.edu

Print on Demand 2018 – Gedruckt auf FSC-zertifiziertem Papier

ISSN 1613-4214
ISBN 978-3-7315-0743-7
DOI 10.5445/KSP/1000077900

Mobile laser scanning based determination of railway network topology and branching direction on turnouts

Zur Erlangung des akademischen Grades
Doktor der Ingenieurwissenschaften
der KIT-Fakultät für Maschinenbau
Karlsruher Institut für Technologie (KIT)

genehmigte
Dissertation
von

DIPL.-INF. DENIS STEIN

Tag der mündlichen Prüfung: 24. Oktober 2017
Hauptreferent: Prof. Dr.-Ing. Christoph Stiller
Korreferent: Prof. Dr.-Ing. Dirk Abel

Abstract

GNSS-based state of the art systems can provide a train-borne localization in many cases. Nevertheless, their positioning accuracy is insufficient when other tracks are nearby. In addition, tunnels for example completely block satellite signals, so that their measurements are not always available. However, ambiguous situations primarily arise on turnouts, which connect several tracks. Thus, a reliable detection of both is the key to success.
For this purpose, a comprehensive detection system is proposed herein. The first part introduces a train-borne setup which perceives additional information on the nearby environment only from a 2d lidar sensor as an independent and diverse sensor principle. In the main part, a multistage method is proposed. The structure of the approach highly resembles that of a common railway network. Based on the expected measurements of an a priori known rail profile, rails and tracks can be detected. In contrast to most related work, turnouts are explicitly considered. Finally, topology and branching events, that provide information on tracks within the nearby environment (topology of the railway network) and branching maneuvers of the railway vehicle on single turnouts (reference position, direction, and side taken thereon), can be raised. The qualitative and quantitative evaluation on a demanding test ground with varied topologies and a large number of branching situations shows a high level of correctness and completeness of the detections. Moreover, the reference position on single turnouts can be determined with an accuracy of one meter. The last part draws conclusions from this work and shows in which ways it can be extended. At the same time, it identifies the required information for the integration of those events into a train-borne localization system which even considers the potential inaccuracy

and incompleteness. The prototypically realized integration into a train-borne localization system was already able to reduce or even eliminate negative effects, such as ambiguity problems, whereby the individual benefits depend on the topological situation.
In summary, this work demonstrates that rails, tracks, and turnouts can be detected systematically. Furthermore, the approach provides valuable information on the environment and in particular the driven route, even if GNSS measurements are not available. It allows to improve the estimated position of a railway vehicle and to assess several possibilities independently. Overall, it shows that detections from 2d lidar sensor measurements can even reduce ambiguity problems in train-borne localization systems.

Kurzfassung

Bereits heute ist eine rein fahrzeuggebundene Lokalisierung von Schienenfahrzeugen mit satellitenbasierten Systemen in vielen Fällen möglich. Jedoch ist deren Genauigkeit oft unzureichend. Dies gilt insbesondere für Orte, in deren Nähe sich parallele Gleise befinden. Zudem kann die notwendige Sichtbeziehung zu Satelliten unterbrochen sein, was beispielsweise in Tunneln der Fall ist. Deshalb ist eine Positionsbestimmung nicht unter allen Umständen möglich. Mehrdeutige Situationen entstehen vor allem durch das Befahren von Weichen, also an Orten, an denen ein Wechsel zwischen mehreren Gleisen möglich ist. Um eventuell auftretende Mehrdeutigkeiten bereits bei ihrer Entstehung erkennen zu können, wird in dieser Arbeit die Detektion von Nachbargleisen und Weichen verfolgt.

Diese Arbeit schlägt dazu ein umfassendes Lösungspaket vor, welches ein etabliertes Setup um ein komplementäres Messprinzip erweitert (Laser-Abstandsmessung). Mit Hilfe eines Einzeilen-Laserscanners können Informationen über das nähere Umfeld des Schienenfahrzeugs erfasst werden. Dazu wird ein mehrstufiges Detektionsverfahren beschrieben, das ausschließlich auf zum Fahren notwendige und damit stets vorhandene Infrastruktur abzielt. Die Struktur des Verfahrens spiegelt dabei den Aufbau eines Gleisnetzes wider. Die wiederholte Beobachtung von Schienenprofilen in den Messdaten ermöglicht die Erkennung von Schienen- und Gleisverläufen während der Fahrt. Daraus können Informationen zur Existenz von Nachbargleisen (Topologie des Gleisnetzes) sowie zu Weichen und deren Befahrrichtung (Referenzposition, Richtung sowie befahrene Seite) detektiert werden, die als Events bereit gestellt werden. Messfahrten in einem Testgelände ermöglichten eine qualitative und quantitative Evaluierung des vorgeschlagenen Verfahrens. Trotz anspruchsvoller und vielfältiger Topologie und einer Vielzahl an Weichenbefahrungen zeichnen sich die Ergebnisse durch ein hohes Maß an Korrektheit und Vollständigkeit aus. Während viele verwandte Arbeiten Weichen nicht berücksichtigen, ermöglicht diese Arbeit die Bestimmung ihrer Referenzposition mit einer Genauigkeit von einem Meter. Abschließend werden die Ergebnisse dieser Arbeit zusammengefasst und ausgewertet sowie Erweiterungsmöglichkeiten aufgezeigt. Zudem werden die erforderlichen Informationen zur Integration der Detektionen in ein bordautonomes Lokalisierungssystem identifiziert und es wird gezeigt, wie dabei selbst mögliche Ungenauigkeiten oder Unvollständigkeiten berücksichtigt werden können. Bereits eine prototypische Integration der vorgeschlagenen Events in ein Lokalisierungssystem demonstriert, dass negative Auswirkungen, zum Beispiel in Form von Mehrdeutigkeiten, reduziert oder gar komplett eliminiert werden können.

Diese Arbeit zeigt, dass Schienen, Gleise und Weichen systematisch erkannt werden können. Zudem funktioniert das Verfahren selbst dann, wenn eine satellitenbasierte Lokalisierung gestört oder gar nicht möglich ist. Zusätzlich liefert es wertvolle Informationen über das nähere Umfeld des Schienenfahrzeugs und insbesondere seine Fahrt durch das Gleisnetz. Diese erlauben es, die geschätzte Fahrzeugposition in einem Lokalisierungssystem fortlaufend zu aktualisieren und im Falle mehrerer Möglichkeiten, diese individuell zu bewerten. Insgesamt wird gezeigt, dass bereits Detektionen aus Messungen eines Einzeilen-Laserscanners ausreichen, um Mehrdeutigkeiten in der bordautonomen Lokalisierung zu reduzieren.

Acknowledgements

This thesis was developed during my activity as a research assistant at the Institute of Measurement and Control Systems of the Karlsruhe Institute of Technology and at the FZI Research Center for Information Technology.
I am extremely grateful to Prof. Dr. Christoph Stiller for supervising this work. He granted me the freedom to act on my own responsibility and, at the same time, provided me financial security. I sincerely thank Prof. Dr. Dirk Abel for acting as co-advisor and in this way demonstrating his interest in the matter.
Following an old tradition, I wish to thank all colleagues, who supported me in a number of ways, so that I was able to improve my outcomes. I am infinitely grateful to Dr. Martin Lauer. He introduced me to the railway domain, always had a positive and constructive influence on my work, and was an indispensable mentor. Furthermore, special thanks go to Max Spindler and Ö. Şahin Taş, with whom I spent a lot of time together, not only discussing theories and methods. For substantive suggestions and insights into existing software, my thanks goes to Dr. Julius Ziegler, Dr. Tobias Strauß, and Claudio Bandera. For the excellent support in administrative and technical issues, I sincerely thank Sieglinde Klimesch, Erna Nagler, Werner Paal, Goran Cicak, and Günter Barth.
For their numerous inspirations and contributions, I would like to thank my students Z. Málna Pólya, Timo Hackel, Benedikt J. Riehm, Kevin Daiß, Janina Kuper, Elmar Adam, Daniel Szeifert, and especially Sören Pottberg.
For quickly and straightforwardly provided lidar sensors and deeper insights into their functionality I sincerely thank Ralf Rössling and Julian-Robin Krato from Pepperl+Fuchs. Thanks go to the Institute of Rail Vehicles and Transport Systems at RWTH Aachen University, especially Florian Eßer, Anselm Daniel,

Daniel Lüdicke, and Andreas Schütz, which provided the opportunity to use their experimental railway vehicle "IFS1" for measurement runs.
Special thank goes to Berit Lochner for assistance in graphic design. Additionally, I want to thank her and Bettina Kirchner for final proofreading.
Finally, I want to express my great gratitude to my parents and more particularly to my long-standing girlfriend, which supported me patiently, assisted me in difficult times, and gave me strength and motivation. This work would not have been possible without these people.

Karlsruhe, June 2017 *Denis Stein*

Contents

List of abbreviations . . . . **xi**

1 Introduction **1**

1.1 Motivation 1

1.2 Problem statement 3

1.3 Solution approach 5

1.4 Aims, contributions, and outline 7

2 Method **9**

2.1 Overall concept 9

2.2 Setup, sensor, and data 12

2.2.1 Mobile perception system 12

2.2.2 Lidar sensor data 14

2.3 Detection of rail profiles in single scans 16

2.3.1 Restriction of the evaluation range 16

2.3.2 Detection of extensive structures 20

2.3.3 Detection of key points for rails 24

2.3.4 Template matching 28

2.4 Detection of rails and tracks within consecutive scans 32

2.4.1 Detection of rail sections 32

2.4.2 Detection of track sections 36

2.5 Inference from detected rails and tracks 42

2.5.1 Detection of topology events 43

2.5.2 Detection of the branching direction on turnouts 48

2.6 Interim Conclusion 56

3 Experimental results **57**
3.1 Overview of the experiments 57
3.1.1 Experimental setup 58
3.1.2 Test ground 59
3.1.3 Data sets 60
3.2 Exemplary evaluation 63
3.2.1 Detection of rail profiles 64
3.2.2 Detection of rail sections 66
3.2.3 Detection of track sections 68
3.2.4 Detection of topology events 70
3.2.5 Detection of the branching direction on turnouts 72
3.3 Quantitative evaluation 75
3.3.1 Rail profile, rail section, and track section detection . . . 76
3.3.2 Detection of topology events 80
3.3.3 Detection of the branching direction on turnouts 83

4 Discussion **93**
4.1 Review of the proposed approach 93
4.1.1 Remaining limitations and their elimination 93
4.1.2 Conclusions from the results 96
4.1.3 Adaptability to varied environments 98
4.1.4 Integration into a self-localization framework 104
4.2 Review of the state of the art 111
4.2.1 Setup, sensor, and data 113
4.2.2 Method 115
4.2.3 Results and evaluation 119
4.2.4 Interim conclusion 120

5 Summary, conclusions, and future work **121**
5.1 Summary 121
5.2 Conclusions 122
5.3 Future work 123

References **125**

Publications by the author **135**

Supervised theses **137**

List of abbreviations

2d/3d	**2/3-d**imensional
ALS	**a**irborne **l**aser **s**canning
BOStrab	**Stra**ßen**b**ahn-**B**au- und Betriebs**o**rdnung (German light railway regulations)
CBTC	**c**ommunication-**b**ased **t**rain **c**ontrol
DIN	**D**eutsches **I**nstitut für **N**ormung (German Institute for Standardization)
EBO	**E**isenbahn-**B**au- und Betriebs**o**rdnung (German railway regulations)
ECS	**e**ddy **c**urrent **s**ensor
EGNOS	**E**uropean **g**eostationary **n**avigation **o**verlay **s**ervice
EN	**E**uropäische **N**orm (European standard)
ESBO	**E**isenbahn-**B**au- und Betriebs**o**rdnung für **S**chmalspurbahnen (German narrow-gauge railway regulations)
FL	**f**acing **l**eft
FN	**f**alse **n**egative
FP	**f**alse **p**ositive
FR	**f**acing **r**ight
GaLoROI	**Ga**lileo **lo**calisation for **r**ailway **o**peration **i**nnovation
GLONASS	**glo**balnaya **na**vigatsionnaya **s**putnikovaya **s**istema (Russian GNSS)
GNSS	**g**lobal **n**avigation **s**atellite **s**ystem
GPS	**g**lobal **p**ositioning **s**ystem

ICE	**I**nter**c**ity-**E**xpress
IMU	**i**nertial **m**easurement **u**nit
laser	**l**ight **a**mplification by **s**timulated **e**mission of **r**adiation
lidar	**li**ght **d**etection **a**nd **r**anging
MLS	**m**obile **l**aser **s**canning
PCA	**p**rincipal **c**omponent **a**nalysis
RANSAC	**ran**dom **sa**mple **c**onsensus
RTK	**r**eal **t**ime **k**inematic
TL	**t**railing **l**eft
TLS	**t**errestrial **l**aser **s**canning
TP	**t**rue **p**ositive
TR	**t**railing **r**ight
UIC	**u**nion **i**nternationale des **c**hemins de fer (International Union of Railways)
WAAS	**w**ide **a**rea **a**ugmentation **s**ystem

1 Introduction

Automation is becoming increasingly more important for rail transport. It effects railway traffic control and even enables automatic train operation. With the help of modern operation procedures, safety can be ensured, collisions can be avoided, and the capacity of the railway network can be enhanced. For all these cases, accurate positioning is indispensable.

This work contributes to this challenge by improving track-selective localization. It is shown which specific requirements arise from the characteristics of a railway scenario and it is furthermore illustrated in which way the accuracy of railway vehicle self-localization can be improved effectively.

1.1 Motivation

A railway vehicle can carry large amounts of goods or several hundred passengers at high velocities over long distances. However, the characteristics of the system differ significantly from those of road traffic. Railway vehicles can normally not be driven on sight, since the static friction is about eight times smaller than for road vehicles, which results in very long braking distances [Pachl 2016]. Furthermore, railway vehicles are guided. They can thus not avoid an obstacle and change their track only on turnouts.

As an illustration, the scenario in Figure 1.1 should be considered. Railway vehicle v_1 is scheduled to arrive on platform p_1. If it arrives on another platform like p_2 for operational reasons, passengers might be confused or frustrated at the best. However, if another railway vehicle v_2 comes along, dangerous situations can arise, such as a potential frontal or lateral collision. Then, the second railway vehicle v_2 must not enter the track section in front of platform p_2 as long as v_1 has not left it completely. Thus, v_2 has to wait at its current position as shown in Figure 1.1. In contrast, if v_1 is located at platform p_1 as planned, v_2 can drive towards platform p_2, since both railway vehicles do not come into conflict.

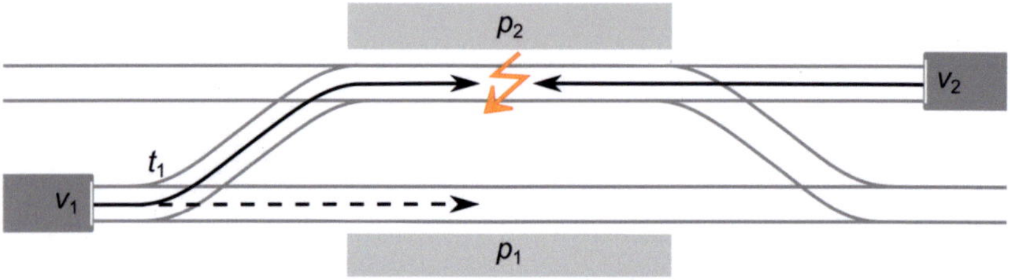

Figure 1.1: Exemplary railway scenario at a station. Arrows indicate the (intended) driving directions of both railway vehicles v_1 and v_2 towards platforms p_1 and p_2. t_1 denotes a single turnout.

To prevent these dangerous situations, the locations and directions of the railway vehicles need to be known. The location includes the knowledge of the lateral and the longitudinal position (i.e., an identifier of the track and the position thereon). The distance traveled as the only information is insufficient for this, because it does not allow to differentiate between whether railway vehicle v_1 is located on the track section in front of platform p_1 or the parallel and nearby track section in front of platform p_2 after passing turnout t_1 in Figure 1.1. However, these ambiguity problems can only arise on turnouts, so that three solutions are conceivable for this. A human observer could be placed near turnout t_1 and monitor all passing railway vehicles. However, this might be necessary around the clock, requires permanent concentration, and is resource intensive, since every turnout requires such an observer. Moreover, human factors already are a main cause of railway accidents [Federal Railroad Administration 2017]. The second option is a technical solution of the observer, where additional elements are installed on the railway network, which is the current state of the art. They allow to detect the presence of railway vehicles or exchange information with them. Examples include track circuits, axle counters, balises, and further components of CBTC (communication-based train control)

systems. On the one hand, the concentration problem is eliminated and a high positioning accuracy is enabled particularly by the use of balises. On the other hand, the supplementary infrastructure causes further costs for installation and maintenance. This is not cost-efficient in particular for low density lines, which is why they are often not installed. Thus, the most promising solution is the equipment of railway vehicles with appropriate sensors, which determine the track and the position thereon. This is especially attractive, since the total number of railway vehicles is typically much smaller than the number of sites which need to be equipped, such as turnouts[1].
Common solutions without any track-side installation and their main challenges for the railway vehicle position determination are identified subsequently[2].

1.2 Problem statement

As motivated before, the position of a railway vehicle within a railway network should only be determined with information that is available on-board, but does not originate from track-side installations. Such self-localization systems of the current state of research are almost exclusively based on global positions from GNSS (global navigation satellite system)[3], information on the movement of the railway vehicle[4], and a digital map[5] of the railway network. They are surveyed for example in [Marais et al. 2017; Winter et al. 2017].

[1] In addition, physical balises can be replaced by virtual balises [Röver et al. 1998; Lauer and Stein 2015].

[2] Train protection systems also address further aspects, such as the communication with a control station or other railway vehicles, an examination of train integrity and speed limits, or the exchange of information with an interlocking. These are however not within the scope of this work.

[3] This includes current and future global systems, such as the US-American GPS, the Russian GLONASS, the European Galileo, and the Chinese BeiDou, as well as satellite or ground based augmentation systems and enhancement techniques, such as the US-American WAAS, the European EGNOS, Differential GPS, or RTK [Bacci et al. 2012].

[4] This includes the traveled distance, velocity, or acceleration, which are measured by an odometer, tachometer, or accelerometer respectively.

[5] This digital map contains information on the geometry and geography of the railway network and its topology, where various connection types, such as single turnouts, are used.

GNSS-based systems allow continuous positioning in many cases and can increase the safety level, since they would also work in case of an error, such as a wrongly set route on a turnout. The availability, accuracy, and reliability of the used sensors is however insufficient in some cases. The error of odometers or tachometers increases with the traveled distance. GNSS receivers suffer from signal shadowing in tunnels or under bridges, multi-path effects near buildings, and further disturbances or interferences [Saab 2000b; Stadlmann et al. 2010; Heirich et al. 2012; Allotta et al. 2014; Marais et al. 2017]. Those effects result, inter alia, in an inadequate lateral resolution, especially where nearby tracks exists. Thus, the track which the railway vehicle is currently running on can only be determined *ambiguously*, which is often referred to as missing *track-selectivity*. With regard to the continuous position determination, [Winter et al. 2017] identified the unsolved *overall problem*:

> How can the location of the railway vehicle as well as its change from one track to another be determined reliably regardless of environmental conditions?[6]

Previous research showed that the track can be determined unambiguously only after driving a few hundred meters, especially when there are parallel tracks nearby [Lauer and Stein 2015]. However, areas with reduced or no signal availability can always occur. Thus, a further improvement of the GNSS-based localization itself is not expedient for this. A promising alternative is the integration of another sensor principle that is independent from the already integrated GNSS and movement measurements [Mirabadi et al. 1996]. In this way the thereof approximately determined position of the railway vehicle can be further refined. This leads to the *specific problem* that is addressed herein:

[6] This includes both the self-localization after an initialization of the system as well as the further continuous determination of the railway vehicle position.

Which sensor can provide additional information in as many situations as possible, which allows to reduce or eliminate the previously mentioned ambiguity problems?

Potential solutions are considered in the subsequent section.

1.3 Solution approach

As motivated before, a sensor is required that is preferably independent from most environmental conditions, that does not suffer from the drawbacks of GNSS-based systems, and whose acquisition costs are admissible for a railway vehicle[7]. Various ideas have been proposed for a solution, which include the detection of stationary objects (e.g., bridges, level crossings, masts, platforms, or tunnels [Heirich et al. 2012; Rahmig et al. 2013; Daoust et al. 2016]) or location-dependent properties (e.g., changes of the curvature or superelevation of the track as well as inductance, magnetic field, or vibration [Saab 2000a; Heirich et al. 2011; Heirich et al. 2013a; Heirich and Siebler 2015; Spindler et al. 2016b]). For this, typical additional on-board sensors include:

- inertial measurement units (IMU) [Saab 2000a; Plan 2003; Broquetas et al. 2012; Heirich et al. 2013b; Heirich et al. 2013a],
- cameras [Maire 2007; Kaleli and Akgul 2009; Wohlfeil 2011; Qi et al. 2013; Corsino Espino and Stanciulescu 2013; Zwemer et al. 2015],
- light detection and ranging (lidar) sensors [Blug et al. 2004; Rahmig et al. 2013; Yang and Fang 2014; Hackel et al. 2015], and
- eddy current sensors (ECS) [Mesch et al. 2000; Geistler and Böhringer 2004; Hensel et al. 2011].

[7] [Manz et al. 2015] for example, assume that the acquisition costs for the whole train-borne localization system which consists of four sensors and the corresponding processing units are not more than 50,000 EUR. Thus, they are assumed to be less than 10,000 EUR per sensor herein.

However, with regard to the considered problem, two observations are particularly striking:

1. Since railway vehicles are guided, track-selectivity only gets lost when the railway vehicle *passes a turnout* in facing direction. In such cases, one track splits up into typically two tracks (cp. for example railway vehicle v_1 in Figure 1.1, which can run either towards platform p_1 or p_2 when passing turnout t_1 in reading direction).

2. In addition, ambiguous situations result from the fact that at least *one more track* is a possible location for the railway vehicle. Resulting from the first observation, the other track is located in close proximity directly after passing the turnout.

Thus, only the detection of nearby tracks and turnout passings addresses the cause of the ambiguity problem [Mesch et al. 2000; Stein et al. 2014a].
IMU and ECS can only provide information on the running of the railway vehicle on the ego track, but cannot include its environment or enable a look ahead. Furthermore, ECS is still a research subject and thus not close to the market. In contrast, different versions of IMUs are available, but they either strongly suffer from drift or are quite expensive [Heirich et al. 2012][8].
On the one hand, cameras and lidar sensors can perceive the environment, whereby the former ones provide a higher resolution. On the other hand, their measurements might be affected by environmental conditions, such as illumination and weather (e.g., rain, fog, or snow). Cameras in particular depend on external illumination[9] and require an interpretation of the measured structures, while lidar sensors provide their own illumination and measure distances directly and precisely [Reiterer et al. 2014].

[8] More stable sensors start at 15,000 EUR and can exceed 100,000 EUR.

[9] The lighting of the railway vehicle is insufficient for this, for example when passing a tunnel or when driving at night.

Thus, lidar sensors appear to be the most promising sensor principle for a spatially close detection of the above mentioned elements and will be used herein.

1.4 Aims, contributions, and outline

This work investigates how additional information on the environment can be perceived by a low cost lidar sensor that is mounted on a railway vehicle. This is often called mobile laser scanning, whereby laser is an acronym for light amplification by stimulated emission of radiation. The lidar sensor extends the common train-borne GNSS-based localization setup, which suffers from missing track-selectivity. The proposed approach focuses on the detection of omnipresent infrastructure elements, which are necessary for the running of the railway vehicle. In doing so, the following issues are addressed in particular:

- Which setup is beneficial for this problem? Which characteristics of the lidar sensor are crucial and which information can be used?
- How can rails and tracks[10] be detected in mobile lidar sensor measurements?
- How can topological information on the nearby environment of the railway vehicle and in particular the route taken within the railway network be determined?
- Which characteristics of these detections are valuable for evaluating position ambiguities and are thus required for their integration into a train-borne localization approach?

Chapter 2 presents the overall idea of the proposed approach and gives an overview of basic railway characteristics. Based on this, a universal setup is introduced. In contrast to most related work, the proposed approach makes

[10] This includes the ego track, i.e., the track, on which the railway vehicle is running, as well as parallel and diagonal tracks in the nearby environment.

use of all information from a 2d lidar sensor, which includes distance and echo measurements as well as the knowledge of their temporal and spatial arrangement. Furthermore, it describes, how rails and tracks can be detected from mobile lidar sensor measurements methodically. It makes use of different features such as occluding edges and combines proven region based as well as model based methods. Moreover, this approach explicitly considers turnouts and their general structure, but it does not require any pattern database for their detection. In this way information on nearby tracks, positions of turnouts on the ego track, and branching directions thereon can be derived, which directly addresses the causes of ambiguity problems. In addition, the approach even works when GNSS or movement measurements are disturbed or unavailable, since it only requires lidar sensor data. A comprehensive experimental evaluation on a demanding railway test ground provides a variety of branching situations and diverse topologies. Chapter 3 evaluates the obtained results qualitatively and quantitatively. Based on this, Chapter 4 extensively discusses the strengths and weaknesses of the proposed approach, highlights the contributions of this work, and shows extension opportunities. Furthermore, the results have already been integrated into a train-borne localization system. Since the position of turnouts as well as the branching direction thereon is known, positions can be adjusted and potential ambiguities further assessed. The final chapter summarizes this work, draws overall conclusions, and provides an outlook on future research directions.

2 Method

In order to detect the topology of the railway network and the branching direction on turnouts, a multistage approach is proposed that considers the structure of a common railway network and its characteristics. The motivation for the overall concept is illustrated at first. Afterwards, a train-borne setup with an additional lidar sensor is introduced, which provides a series of vertical cross-sections of the nearby environment. A multistage approach is presented in the main part. It detects rail profiles within single scans using different features and model based approaches. The spatial clustering of rail profiles over consecutive scans allows to recognize rail and track sections. This way information on the topology of the railway network can be inferred. Finally, positions of turnouts on the ego track as well as the direction in which the railway vehicle passed them can be derived.

2.1 Overall concept

Since railway vehicles are guided, their infrastructure requires well defined characteristics. Thus, the basic idea is to analyze the structure of a common railway network top-down. A special focus is on those elements which are most important for the running of a railway vehicle. Based on this general decomposition, a bottom-up multistage approach is proposed which resembles a railway network and complies with the structure of this chapter. The particular characteristics of those elements are described more detailed with the relevant detection steps, while this section provides a clear overview.

(a) Two parallel tracks that are connected by two single turnouts.

(b) Several tracks with a diamond crossing in front of the train.

Figure 2.1: Exemplary railway scenarios. Several tracks and different connection types, such as single turnouts and a diamond crossing, are shown. Each track is built up of two parallel rails with a fixed distance (photos by courtesy of Martin Lauer).

The *railway network* consists of a set of *tracks* (cp. Figure 2.1). These tracks can be arranged in different ways, e.g., parallel or intersecting, and at varying distances to each other. They are primarily interconnected by turnouts[11] (cp.

[11] To be consistent, the generic term "turnout" is used for the railway network element that allows branching. Other common terms, in particular "point switch" or "(set of) points", are in conflict especially with the measurements of the lidar sensor (point) or their arrangement (point cloud). For the latter ones, the generic term "measurement" is used herein. However, it cannot be avoided that common railway terms contain "point" or "switch".

Figure 2.1a) and diamond crossings[12] (cp. Figure 2.1b). It is essential that only turnouts allow the railway vehicle to change from one track to another. In contrast, two tracks intersect at diamond crossings where turning is not possible. Tracks consist of a tuple, normally a pair, of parallel counterparts called *rails*. Finally, those rails have a certain *profile*. This decomposition can also be seen in the left side of Figure 2.2.

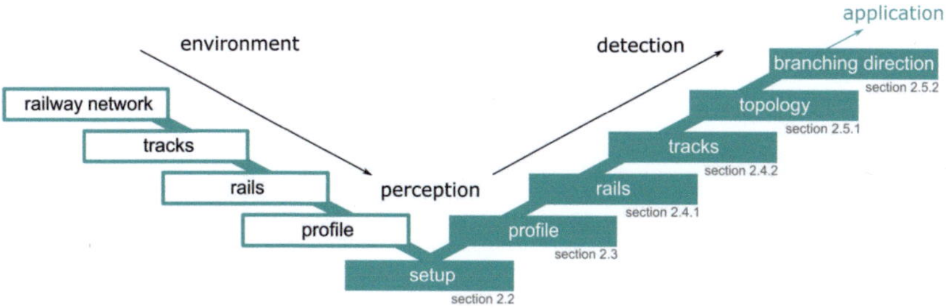

Figure 2.2: Overall concept. Structure of the railway network (left side) and the proposed detection approach (right side).

The proposed multistage approach exactly follows the structure of the environment, but in reversed order (cp. the right side of Figure 2.2). It takes advantage of the fact that all elements which are fundamental for the running of the railway vehicle are usually built up of rails and are available in almost every scenario. Based on the detection of their *profile*, sections of *rails* and *tracks* can be determined. Their arrangement provides valuable information on the *topology* of the nearby environment. Finally, the location of turnouts and the route taken thereon can be inferred, which is referred to as *branching direction*. This information can be used in an application, such as a train-borne localization system.

[12] The common terminology might be confusing. Different variants of both connectors, i.e., turnouts and diamond crossings, can contain the words "diamond crossing". The connection type "diamond crossing" is comparable to four way intersections where turning is prohibited. In contrast, "diamond crossings with single or double slips", which belong to the connection type turnout, are built up of a centered diamond crossing and two or four interleaving single turnouts. Thus, they additionally allow a change of the track, i.e., left turns from one (in case of single slips) or two (double slips) ways and right turns vice versa.

In order to detect rail profiles, rails, and tracks as well as the topology and branching direction on turnouts, a *setup* for a mobile perception system with a lidar sensor is proposed subsequently.

2.2 Setup, sensor, and data

The proposed multistage approach requires a setup for the detection of rails and tracks (cp. the lower part of Figure 2.2). Subsequently, it is described which characteristics of a lidar sensor and its mounting are relevant for these detections. Finally, the corresponding measurements are abstracted in such a way that the approach is independent of a specific setup.

2.2.1 Mobile perception system

Throughout this work, a setup with one lidar sensor is considered. It is mounted on a railway vehicle as shown in Figure 2.3. Its laser beam is rotating in one plane. That is why it is called a single-layer[13] lidar sensor. In this way it perceives the shape of the environment in the x-z-plane, whereby the optical center of the lidar sensor is the origin of the 2d Cartesian coordinate system x-z. It usually provides hundreds of thousands distance measurements per second, while all measurements within one revolution of the laser beam are called a scan. The laser beam provides the necessary lighting, so that the sensor is independent from illumination conditions and properly operates even at night.
Lidar distance measurements from such a moving ground vehicle are also referred to as MLS (mobile laser scanning) application. However, the same location is typically measured just once, while the railway vehicle moves (cp. Figure 2.3c).

[13] The proposed approach can also be applied to each of several layers stacked over each other in a multi-layer lidar sensor. Since it provides additional cross-sections and a forecast in y-direction, objects might be detected slightly earlier. However, multi-layer sensors are still too expensive and typically provide a coarser angular resolution. Thus, they are not considered in the following.

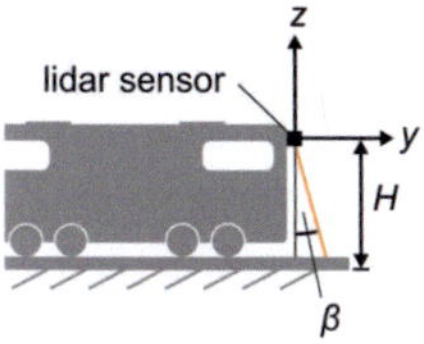

(a) Side view.

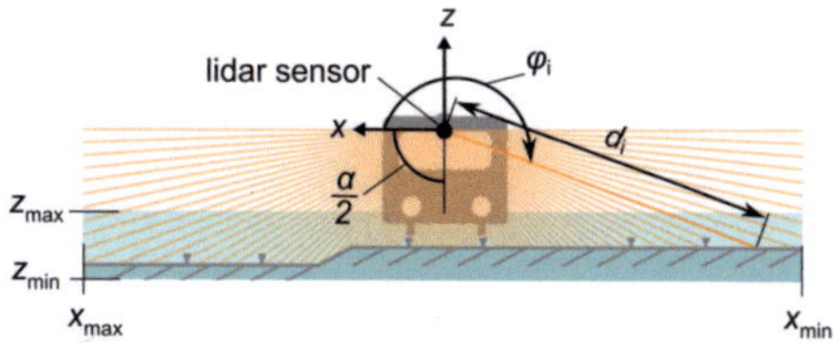

(b) Front view.

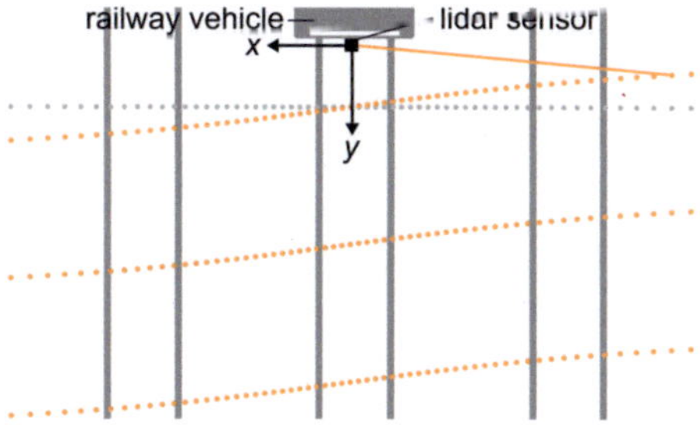

(c) Top view.

Figure 2.3: Proposed setup with a lidar sensor on the front of a railway vehicle. Exemplary measurements are shown in orange (orientation of the three dimensions x, y, z, mounting height H, inclination angle β, field of view α, radial distance d_i, measurement angle φ_i, boundaries of the measuring range $x_{\min}, x_{\max}, z_{\min}, z_{\max}$). The spatial distribution of the measurements on a ground plane is shown by orange dots in (c), while constant railway vehicle velocity is assumed. Gray dots represent exemplary measurements at a standstill or their projection on a common y-value per scan. The latter representation is used subsequently regardless of the velocity.

Previous research showed that the proposed centered and upright mounting (i.e., inclination angle $\beta = 0°$, cp. Figure 2.3a) benefits from various reasons compared to a tilted setup [Stein et al. 2014a; Stein et al. 2016a][14]. For example, the loss of information caused by reflections at flat incidence angles is reduced. Furthermore, vibrations or the pitching of the railway vehicle result in minimal variations in y-direction. Finally, the protection against environmental influ-

[14] If an upright mounting is not possible, the inclination angle $\beta > 0°$ is used for a transformation. A scaling of all z-values by $\cos(\beta)$ projects them into a cutting plane in x-z-direction. Additionally, the forecast in y-direction of $H \cdot \tan(\beta)$ has to be considered when referencing the observed place.

ences, such as rain or dust accumulating on the sensor housing, can be implemented most effectively.
The mounting height H should be the highest technically feasible point. In this way a better overview also on neighboring tracks is enabled. Furthermore, occlusions by other railway infrastructure elements can be reduced. However, the maximal height for railway vehicles is defined by the structure gauge, which is for example 4.74 m in Germany [EBO].
It is assumed that the lidar sensor at least covers the area bounded by $x_{\min}$, $x_{\max}$, $z_{\min}$ and $z_{\max}$ *below* its mounting (cp. Figure 2.3b). Thus, the lidar sensor has a maximal measurement distance of $d_{\max} \geq \sqrt{x_{\min}^2 + z_{\min}^2}$ and a field of view $\alpha \geq 2 \cdot \arctan(x_{\max}/|z_{\max}|)$ when — without loss of generality — axial symmetry is assumed ($|x_{\min}| = x_{\max}$). [Stein et al. 2016a] already showed that $\alpha = 180^\circ$ (as illustrated in Figure 2.3b) and $d_{\max} = 10\,\mathrm{m}$ are optimal for the investigations herein. Furthermore, an angular resolution of $\Delta\varphi \leq 1/10^\circ$ allows a sufficient discretization of the scanned objects in lateral direction (Figure 2.3 illustrates $\Delta\varphi \approx 2^\circ$), while a measurement rate of $f \geq 50\,\mathrm{scans/s}$ is recommended for the discretization in longitudinal direction.
Further lidar sensor properties and their influence on the measurements are described, e.g., in [Wölfelschneider 2009; Stein et al. 2014a]. [Stein et al. 2016a] give an overview of appropriate products, whereby the most suitable sensor will be used in an experimental evaluation in Chapter 3. Finally, the upcoming section describes the measurements in such a way that the proposed approach is independent of a specific setup.

2.2.2 Lidar sensor data

In the following, the Cartesian coordinate system x-y-z as shown in Figure 2.3 is used. The movement of the railway vehicle, and thus the lidar sensor, in driving direction is represented by y. The measurements of the *uprightly* mounted lidar sensor are described in the lateral and vertical dimensions x and z. The distance between consecutive measurements within one scan is quite small[15]. Thus, the longitudinal motion within each scan is neglected. This is often done in lidar sensor data processing (see also [Kutterer 2010]) and is illustrated by gray dots in the first scan in Figure 2.3c.

Each measurement within a scan can be described by a triple

$$m_i = (\varphi_i, d_i, e_i) \tag{2.1}$$

in polar representation with $1 \leq i \leq N$ (cp. Figure 2.3b). The measurement angle φ_i is discretized equidistantly with the angular resolution $\Delta\varphi = {}^{360°}\!/_{N}$. It starts with $\varphi_1 = 0°$ and covers the interval $[0°; 360°)$. The radial distance d_i (in meter) and the dimensionless echo e_i of the i-th measurement are positive. The polar representation φ_i and d_i can be transferred into a Cartesian description of the shape of the environment

$$\begin{pmatrix} x_i \\ z_i \end{pmatrix} = d_i \cdot \begin{pmatrix} \cos(\varphi_i) \\ \sin(\varphi_i) \end{pmatrix}. \tag{2.2}$$

(x_i, z_i) is a point in the x-z-plane. This Cartesian representation is used especially for illustration purposes.

A series of O scans is called a measurement run ($1 \leq j \leq O$). Each scan

$$s^{(j)} = (n^{(j)}, y^{(j)}, m_1^{(j)}, \ldots, m_N^{(j)}) \tag{2.3}$$

is described by an identifier $n^{(j)}$, the railway vehicle position in driving direction $y^{(j)}$, and N measurements (cp. (2.1)). The time intervals between consec-

[15] Even for highest velocities of 360 km/h, the value of y changes by only 0.4 mm between consecutive measurements when assuming an angular resolution $\Delta\varphi = {}^{1}\!/_{14}°$ and a measurement rate $f = 50\,{}^{\text{scans}}\!/_{\text{s}}$ as used in the experimental evaluation in Chapter 3.

utive measurements and consecutive scans are determined by the measurement rate f, the field of view α, and the angular resolution $\Delta\varphi$[16].
In summary, the representation in (2.1)–(2.3) makes the measurements independent from a specific setup. Thus, the subsequently proposed multistage approach works for all measurements within vertical cross-sections of the nearby environment.

2.3 Detection of rail profiles in single scans

As a first step for the detection of the railway network (cp. the overall structure in Figure 2.2), this approach considers the lateral and vertical dimension of the nearby environment within vertical cross-sections. Since rails are located near the ground, a proper detection area is defined at first and extensive areas are identified where rails are not arising. The detection of rails in single scans is based on their location and appearance within the lidar sensor data. It makes use of different features, such as occluding edges, in the distance and echo measurements which further restrict the search space. Finally, a model based approach determines the locations of the rail profile within each scan.

2.3.1 Restriction of the evaluation range

Since the main scope of this work is the rail based detection of tracks and the branching direction of the railway vehicle on turnouts, their position relative to the lidar sensor is focused on at first. Therefore, the typical structure of the railway environment within vertical cross-sections is analyzed in lateral and vertical direction. Based on this, restrictions on the area where rails typically arise are derived. Figure 2.4 shows an exemplary scan which is the basis for the illustration of all subsequent steps.

[16] Equidistant "temporal" discretization in lateral (angular resolution $\Delta\varphi$) and longitudinal direction (measurement rate f) is assumed herein. However, this implies no further restriction. It is fulfilled for all such lidar sensors and does thus not result in loss of generality.

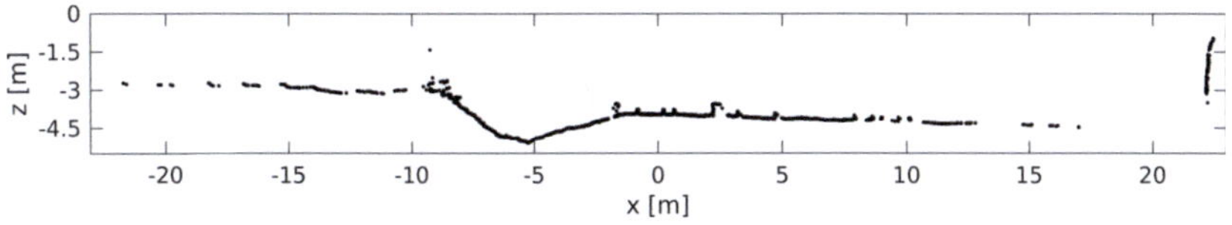

(a) All measurements below the lidar sensor.

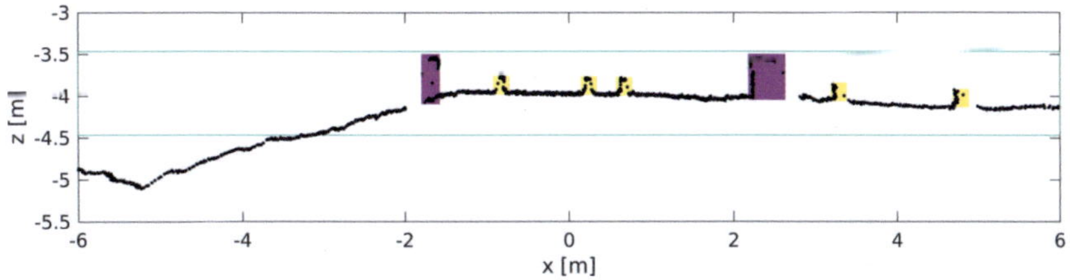

(b) Typical objects within the detection area (boundaries in turquoise): rails (highlighted in yellow) and third rails (purple).

Figure 2.4: Exemplary measurements from a railway vehicle. Also depending on the environment, the gaps between the measurements (black dots) for $|x| \geq 8\,\text{m}$ in (a) increase significantly, while (b) shows a scale-up for $|x| \leq 6\,\text{m}$. Note that a railway vehicle cannot run on third rails, but they can provide power supply to it.

Restriction in lateral direction

Based on the proposed setup and the characteristics of the railway network, the lateral position of the rails of the ego track[17] is expected within the interval $|x| \leq G/2 + x_\text{d}$ where the gauge[18] G is the distance between the rails of a track (as shown in Figure 2.5a) and x_d denotes the maximal deflection of the lateral center of the railway vehicle relative to the center of the track. The maximal deflection x_d takes effect especially in curves and is assumed to be less than 1 m. Typical values for G in Central Europe are 1435 mm (normal gauge) and 1000 mm (meter gauge). The gauge is almost constant, otherwise

[17] The ego track is the track where the considered railway vehicle is running on.

[18] It is defined as the minimal distance 0 to 14 mm below the upper surface of the rail [EN 13848-1; Matthews 2007].

the guided railway vehicle might derail. However, slight variations within the interval $[G-5\,\mathrm{mm}; G+35\,\mathrm{mm}]$ are permissible [Matthews 2007; EBO].

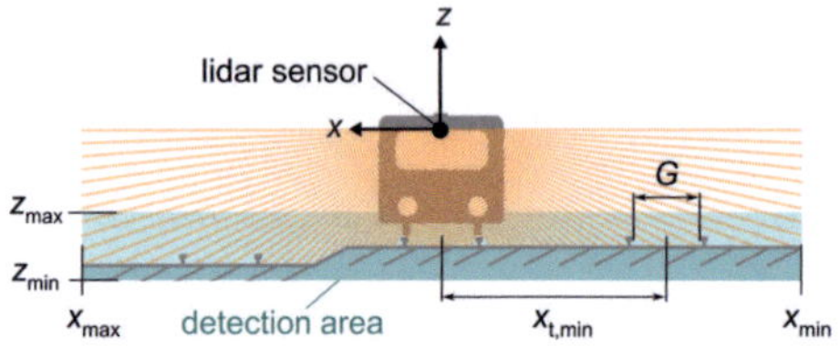

(a) Dimensions of the railway network and the proposed detection area.

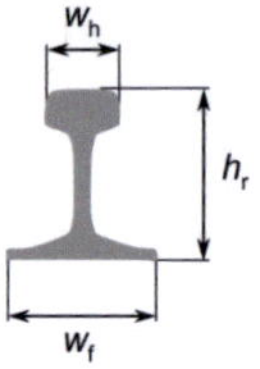

(b) Vignole rail profile.

Figure 2.5: Positions and dimensions of rails and tracks (orientation of the dimensions x, z, boundaries of the detection area $x_{min}, x_{max}, z_{min}, z_{max}$, gauge G, minimal lateral distance of parallel tracks $x_{t,min}$, height of the rail h_r, width of the head of the rail w_h, width of the foot of the rail w_f).

Since this work is also interested in the detection of nearby tracks, their common lateral distance needs to be considered as well. It is defined as the horizontal offset between the center points of both tracks. Its minimum for parallel tracks is 3.5 m in Germany due to historical reasons, but it is typically at least 4 m and also depends on the nearby environment [EBO]. Thus, the leftmost rail of the closest parallel track to the left is expected at $x \approx -x_d - G/2 - x_{t,min}$. For normal gauge tracks with $G \approx 1.4\,\mathrm{m}$ and a minimal lateral track distance of $x_{t,min} \approx 4\,\mathrm{m}$ (as shown in Figure 2.5a), $x_{min} = -6\,\mathrm{m}$ is a reasonable left boundary for the detection area[19]. The same applies to parallel tracks on the right side. Thus, the opposite boundary is $x_{max} = 6\,\mathrm{m}$ (cp. Figure 2.5a).

[19] An extension of the lateral boundary such that two parallel tracks on the left side are covered is not recommended for technical reasons. The leftmost rail is expected at minimal -9 to -10 m to the left, which results in an incidence angle on ground of about 25° even at the largest feasible mounting height $H \approx 4.7\,\mathrm{m}$. This flat incidence angle increases the risk of reflections and occlusions by other objects significantly. Since the ray density on the ground decreases outwards, the rail is discretized worse. Moreover, the second left track would not help to detect the branching direction of the railway vehicle on turnouts, since it is too far away from the ego track. Furthermore, this track has to intersect the track on its right side first. In contrast, [Mikrut et al. 2016] for example limit the lateral area to $\pm 4.75\,\mathrm{m}$.

Finally, corresponding regulations exist for narrow gauge (tracks with $G \leq 1000\,\text{mm}$) [ESBO] or light rail and tram railway networks [BOStrab]. Thus, these considerations apply almost unchanged for them.

Restriction in vertical direction

In order to detect rails and tracks in lidar sensor measurements, the detection area can also be restricted on relevant parts in vertical direction. In this way the search space can be further limited. The vertical distance from the ground is similar to the mounting height H for the ego track. Other tracks are not necessarily on the same vertical position (cp. left and centered track in Figure 2.5a).

Table 2.1: Main dimensions of typical Vignole rails (height of the rail h_r, width of the head of the rail w_h, width of the foot of the rail w_f as shown in Figure 2.5b) [EN 13674-1].

profile	h_r [mm]	w_h [mm]	w_f [mm]
49E1 (DIN S49)	149	67	125
54E3 (DIN S54)	154	67	125
60E1 (UIC 60)	172	72	150

However, the ego track and all tracks merging into or diverging from the ego track, which are later on relevant for the detection of the branching direction, are on almost the same altitude. Otherwise, the railway vehicle could not branch from one track into another. Tracks are elongated straight or curved objects and are typically built up of two parallel rails with a certain profile. Vignole rails are the most common class and are thus considered herein. They are raised objects and their predominant part is typically above the ground (cp. their profile in Figure 2.5b and dimensions of typical variants in Table 2.1[20]). Thus, the evaluation

[20] This also holds for lighter versions of Vignole rails [EN 13674-4]. In contrast, the predominant part of grooved rails [EN 14811] that are typical for tramway scenarios or used on level crossings is not necessarily above ground. Since they only cover a small percentage of the railway network rails, they are not considered herein.

range can also be restricted relative to the mounting height H in vertical direction. An offset of about thrice the mean rail height h_r is a good compromise, which leads to $z_{min} = -H - 0.5\,m$ and $z_{max} = -H + 0.5\,m$ (cp. Figure 2.5a). As a result, only a subset of the measurements below the sensor with $180^\circ < \varphi_i < 360^\circ$ is considered (cp. Figure 2.4b). Note that the detection area cannot cover all tracks that are somewhere in the surroundings of the railway vehicle, e.g., in case of a large number of tracks, such as in front of a railway station (cp. Figure 2.6). However, tracks that merge into the ego track or branch from the ego track always are within the detection area, since they end or start laterally and vertically close to the ego track. Only those are required for the detection of the branching direction of the railway vehicle on turnouts later on.

Figure 2.6: Exemplary scenario near a railway station with several tracks and single turnouts. The parallel tracks left and right of the railway vehicle will be within the detection area. Especially the tracks in the right part of the picture are too far away (photo by courtesy of Martin Lauer).

2.3.2 Detection of extensive structures

In order to further restrict the search space, areas are identified where rails are typically not present. The most relevant area is the ground below the rail foot, which is considered firstly. Secondly, vertical objects are detected.

Ground detection

Idea: The most important area for ground detection is located between the rails of the same track. Its surface depends on the type of the superstructure and consists of different materials, e.g., ballast, concrete, wood, or metal [Matthews 2007]. However, it is flat most of the time and possibly has a small cross slope (cp. also Figure 2.7). Thus, inspired by [Neubert et al. 2008], the existence of at least one ground plane that most probably belongs to the ego track can be assumed. Therefore, a model for the ground plane and a method are needed that decides on the belonging to this ground plane for every measurement within a scan.

Figure 2.7: Exemplary railway scenario with varying materials. The sleepers of the track on the left side are wooden, whereas those on the center tracks are made of concrete. In addition, ballast and mostly rusted rails can be found, while at most small cross slopes occur (photo by courtesy of Martin Lauer).

Method: Since vertical cross-sections in the 2d x-z-plane are considered and major parts of the environment of the rail are assumed to be planar, a straight line hypothesis

$$z = m \cdot x + n \tag{2.4}$$

with slope m and z-intercept n in Cartesian representation can be used for the ground plane. This line can be expressed in polar representation

$$
\begin{aligned}
\rho_{\text{ground},i} &= \frac{\delta_{\text{ground}}}{\cos(\varphi_i - \alpha_{\text{ground}})} \quad \text{with} \\
\alpha_{\text{ground}} &= \arctan(-1/m), \\
\delta_{\text{ground}} &= n \cdot \sin(\alpha_{\text{ground}}).
\end{aligned} \tag{2.5}
$$

δ_{ground} is the distance between the origin of the x-z-coordinate system and this line, α_{ground} is the angle between the x-axis and the normal of the line, and $\rho_{\text{ground},i}$ is the distance of a point on the line measured at angle φ_i from the origin of the x-z-coordinate system (cp. Figure 2.8). Since the origin of the x-z-plane is located in the optical center of the lidar sensor and each measurement of the lidar sensor is described by the measurement angle φ_i and the associated radial distance d_i, the radial distance between the i-th measurement and the line is

$$
d_{\text{ground},i} = \rho_{\text{ground},i} - d_i. \tag{2.6}
$$

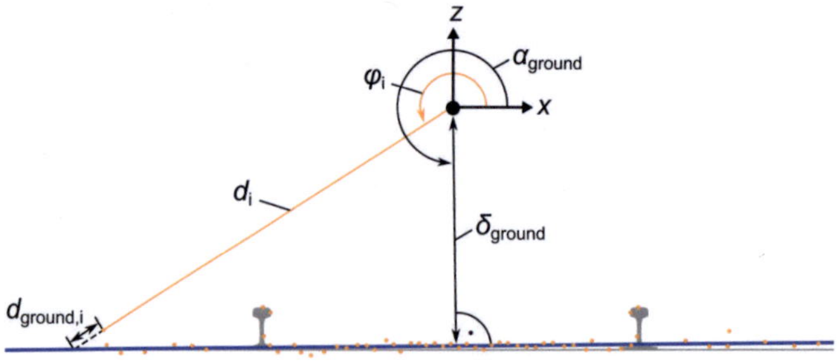

Figure 2.8: Idea of ground detection. Exemplary measurements (orange dots) on two rails and the ground (gray) with the estimated ground plane (dark blue line; orientation of the dimensions x, z, angle between the x-axis and the ground plane α_{ground}, minimal distance of the ground plane from the origin δ_{ground}, radial distance d_i and measurement angle φ_i of the i-th measurement, deviation of the i-th measurement from the ground plane $d_{\text{ground},i}$; the radial distance of the i-th measurement from the ground ρ_{ground} follows (2.6)).

Thus, the binary classifier c_{ground} decides on the belonging of the i-th measurement to the ground plane depending on its distance from the line:

$$c_{\text{ground}}(i) := |d_{\text{ground},i}| \leq \lambda_{\text{ground}}. \tag{2.7}$$

λ_{ground} is the distance threshold for the ground plane.
The idea of the common RANSAC (random sample consensus) estimator, which has been introduced by [Fischler and Bolles 1981], is used, since it is robust against outliers. Two randomly selected points (x_j, z_j) and (x_k, z_k) repeatedly create line hypotheses (2.5) with

$$\begin{aligned} m &= \frac{z_k - z_j}{x_k - x_j} \quad \text{and} \\ n &= z_j - m \cdot x_j. \end{aligned} \tag{2.8}$$

Finally, the line with the highest number of inliers fulfilling (2.7) is selected. Figure 2.9 shows exemplary results of the proposed ground plane detection. Since most measurements on rails are not located on this ground plane, those measurements are not considered as potential key points in the subsequent step. In this way the search space can be restricted effectively.

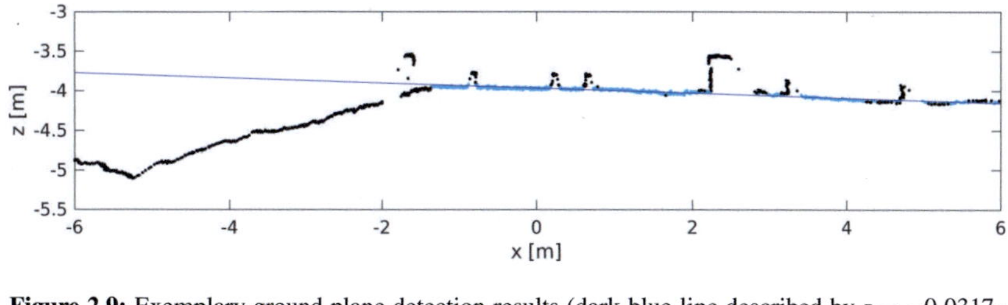

Figure 2.9: Exemplary ground plane detection results (dark blue line described by $z = -0.0317 \cdot x - 3.9670\,\text{m}$). More than 58 % of the 1489 depicted measurements are assigned to the ground plane (light blue dots on ground plane, all others in black), which most probably are not relevant points on the five raised Vignole rails (cp. also Figure 2.4b).

Vertical object detection

Idea: Extensive vertical objects, such as masts of the overhead contact line or signals (cp. Figure 2.1), do also not contain measurements that belong to a rail.
Method: For the detection of vertical objects, the method for ground detection is used with a modified model of (2.5). A vertical line model in polar representation

$$\rho_{\text{vertical},i} = \frac{\delta_{\text{vertical}}}{\cos(\varphi_i)}, \tag{2.9}$$

is used, where δ_{vertical} is the x-value of one randomly selected measurement. Similar to (2.7), the binary classifier for extensive vertical objects c_{vertical} is defined for the i-th measurement of the scan by

$$c_{\text{vertical}}(i) := |\rho_{\text{vertical},i} - d_i| \leq \lambda_{\text{vertical}}. \tag{2.10}$$

It decides whether a measurement is located on the vertical structure or not. $\lambda_{\text{vertical}}$ is used as the distance threshold for vertical objects.
Again, measurements which are located on extensive vertical structures are not considered as potential key point hereafter.
Remarks: Even a potentially insufficient detection of the ground or vertical structures will not influence the correctness of the following steps, but might increase the search space and thus influence the running time, while additional erroneous detections can be filtered out in a later step.

2.3.3 Detection of key points for rails

Most measurements on rails are not located on the previously detected extensive horizontal and vertical structures. Thus, only the remaining areas within the symmetric detection area are further considered. In order to detect rail profiles, the lidar sensor measurements are analyzed for significant features that arise from Vignole rails in single scans. It is subsequently described which of them occur in distance and echo measurements and how they can be detected.

Geometrical rail key point detection

Idea: Vignole rails are raised objects that are located above the ground (cp. Figure 2.5). When they are scanned from a lidar sensor, the part which faces away from the lidar sensor is occluded (cp. Figure 2.10a). This is because the laser beam cannot penetrate materials such as metal. These occlusions are the major geometrical indicator for the existence of Vignole rails in the neighborhood, whereby the subsequently proposed detection method is inspired from [Hackel et al. 2015].

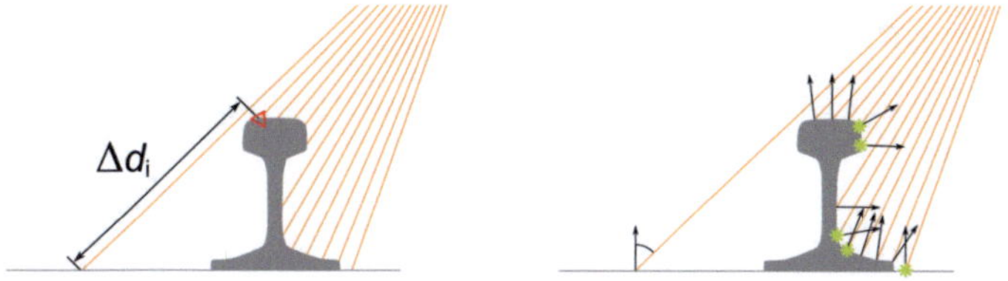

(a) Variation of the measured radial distances.

(b) Variation of the incidence angles.

Figure 2.10: Typical influences used for the key point detection in the x-z-plane (selected laser beams in orange, rail profile in gray). (a) The rail causes an occlusion of the area behind it which results in significant changes Δd_i between neighboring distance measurements (last measurement on the top of the rail is a distance key point; red triangle). (b) The rail shape (surface normals shown as black arrows) causes large changes of the incidence angle (exemplary marked for the left measurement) which results in significant changes of the echo measurement (potential echo key points denoted by green asterisks).

Method: Occlusions result in a significant change of the measured radial distance. These changes are defined by the deviation Δd_i between consecutive measurements:

$$\Delta d_i = \begin{cases} d_{i-1} - d_i & 180^\circ < \varphi_i \leq 270^\circ \\ d_{i+1} - d_i & 270^\circ < \varphi_i < 360^\circ, \end{cases} \tag{2.11}$$

where the innermost measurement is selected. The binary classifier $c_{\text{occlusion}}$ decides whether the deviation of the i-th measurement is significant and thus causes an occlusion:

$$c_{\text{occlusion}}(i) := \Delta d_i > \lambda_{\text{occlusion}}. \tag{2.12}$$

$\lambda_{\text{occlusion}}$ is the corresponding deviation threshold. Measurements fulfilling (2.12) are called an "occluding edge" (cp. Figure 2.11).

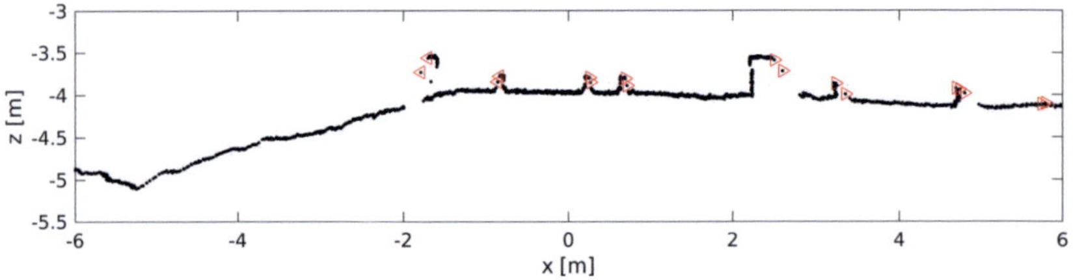

Figure 2.11: Exemplary results of the detection of occluding edges. Occluding edges (red triangles) especially appear at the sides of the rails that face away from the lidar sensor. Black dots represent measurements (cp. also Figure 2.4b).

Remarks: In a smooth, continuous environment, only small changes between consecutive measurements are expected. They range in the size of the measurement noise. Significance is given when this assumption is considerably violated. Thus, $\lambda_{\text{occlusion}}$ should be a small multiple of the standard deviation of the radial distance measurements, which can be determined experimentally.

Echo based rail key point detection

Idea: Besides the geometrical appearance, the reflection is an essential characteristic for the detection of rail profiles. Figure 2.10b shows that the incidence angle of the laser beam changes several times when measuring the rail profile. Furthermore, the material and its reflectivity changes as well, e.g., from ballast or concrete on the left side to blank steel on the top which is caused by the wheel-rail-contact, and rusted steel on the right side of the rail. Both effects result in several significant changes in the reflected energy and especially occur in the neighborhood of rails as already observed by [Blug et al. 2004; Yang and Fang 2014]. Again, a criterion for this effect and thus an indicator for the existence of nearby rails is needed.

Method: The changes in reflection can be observed in the echo measurements. The latter ones are interpreted as ordinal data. Similar to (2.11), the deviation Δe_i between consecutive measurements is defined by

$$\Delta e_i = \begin{cases} e_{i-1} - e_i & 180^\circ < \varphi_i \leq 270^\circ \\ e_{i+1} - e_i & 270^\circ < \varphi_i < 360^\circ. \end{cases} \tag{2.13}$$

The binary classifier c_{echo} decides whether the i-th measurement causes a significant deviation in the echo

$$c_{\text{echo}}(i) := |\Delta e_i| > \lambda_{\text{echo}}. \tag{2.14}$$

λ_{echo} is the threshold for echo deviations. Measurements fulfilling (2.14) are by definition an echo key point (cp. Figure 2.12). Note that upward and downward deviations are considered in contrast to (2.12).

Remarks: In a smooth, continuous environment that consists of homogeneous materials, only smooth changes of the echo are expected. They are near the measurement noise, too. In contrast, when the incidence angle or the material change, significant deviations occur (cp. Figure 2.10b). Again, λ_{echo} should be a small multiple of the experimentally determined standard deviation, but this time of the echo measurements.

The aim of both key point detectors is to further limit the search space. However, missing detections cannot be reconstructed anymore. Thus, Figure 2.11 and Figure 2.12 show more key points than rails. Those additional detections can be filtered out in a later step.

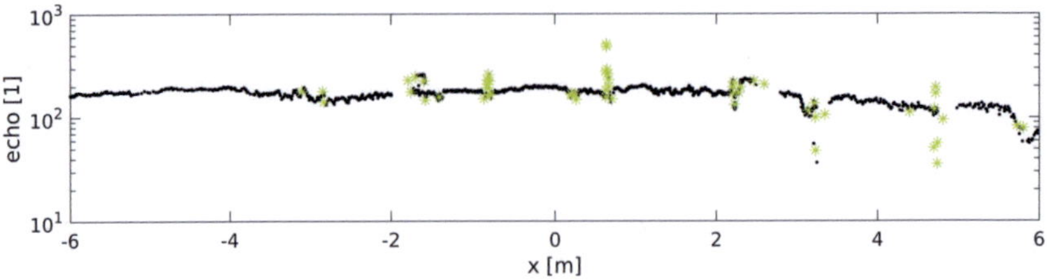

(a) Echo measurements.

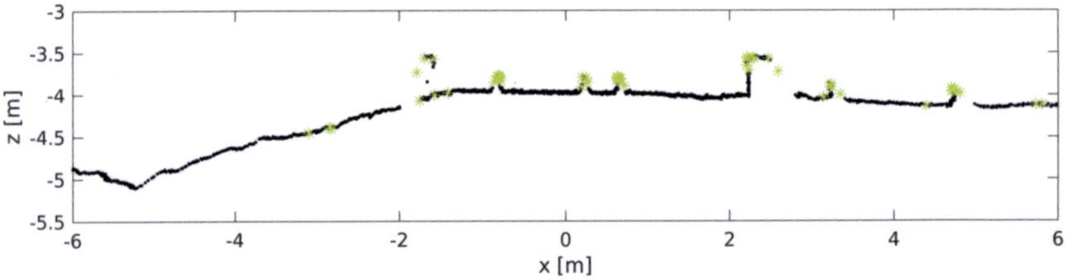

(b) Distance measurements.

Figure 2.12: Exemplary echo key point detection results (represented by green asterisks; black dots represent distance and echo measurements; cp. also Figure 2.4b). From the corresponding distance measurements in (b), it can be clearly seen that echo key points especially appear on parts such as rails where the incidence angle or the material changes frequently.

2.3.4 Template matching

After the steps that focused on the search space, the actual shape of the rail is explicitly considered as proposed in the overall structure in Figure 2.2.

Idea: Rails have a defined geometrical shape which predominantly appears in the x-z-plane and is a priori known (cp. Figure 2.5). This model knowledge can be used to compare their actual measurements to expected ones, since the position of the lidar sensor and the detection area are also known. Thus, a method is required which evaluates whether the rail exists and where it is located within a scan. Several rails which need to be detected can appear per scan.

Method: In order to evaluate the similarity between the rail shape and its appearance in a scan, a template matching approach is used as proposed in [Stein et al. 2016b]. Since the lidar sensor measures distances radially, the templates are also described in polar representation. This allows a direct comparison between actual distance measurements of the lidar sensor and expected ones, which are located on the rail shape model. The structure of the corresponding template library, the search space of each rail investigation, and the use of the template matching for the detection of rail shapes are considered for this.

The template library describes undisturbed measurements within the detection area. It contains all expected measurements $\{\widetilde{\varphi}_i, \widetilde{d}_i\}$ on the rail shape (cp. the magenta dots in Figure 2.13a) for every translation t_x, t_z of the center of the template. $\widetilde{\varphi}_i$ denotes the measurement angle and $\widetilde{d}_i$ is the corresponding radial distance of the i-th measurement in polar representation.

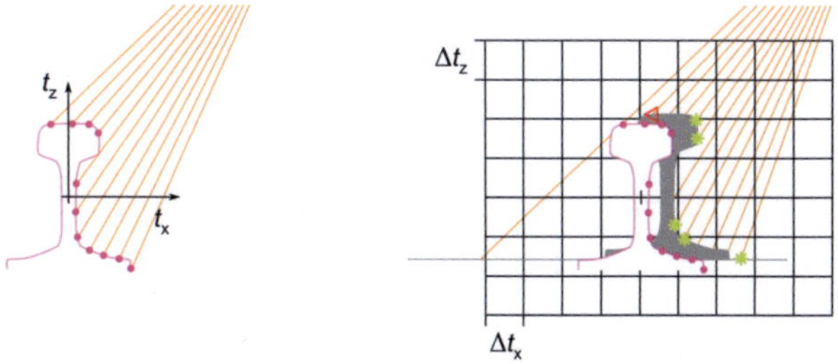

(a) The template library contains all expected measurements (magenta dots) on a shape template (magenta line) for a given translation t_x, t_z.

(b) Exemplary search space for template (magenta) matching on a rail (gray) with previously detected key points (red triangle and green asterisks; cp. also Figure 2.10).

Figure 2.13: Idea of model based rail profile detection in the x-z-plane (selected laser beams in orange). The rectangular search grid in (b) is discretized with the resolution of the translations Δt_x and Δt_z.

Previously detected key points restrict the search space $\mathscr{S}$ in which the template matching applies (cp. Figure 2.13b). Since occluding edges and echo key points are characteristic for lidar sensor measurements of rails, both are required on the template. Thus, all translations in the local neighborhood of an occluding edge (which has been detected on the k-th measurement) are considered if also an echo key point occurs on the template:

$$\begin{aligned}\mathscr{S}_k = \{(t_x, t_z) \,|\, c_{\text{occlusion}}(k) \wedge k \in \mathscr{I} \wedge \exists j \in \mathscr{I} : c_{\text{echo}}(j) \\ \text{with } \mathscr{I} = f_{\text{shape}}(t_x, t_z)\}.\end{aligned} \tag{2.15}$$

f_{shape} provides only indexes of measurements that are located on the considered shape for a given translation t_x, t_z.

In order to allow a direct comparison, the measurement angles used in the template library equal those of the lidar sensor measurements (i.e., $\widetilde{\varphi}_i = \varphi_i$, $1 \leq i \leq N$). Thus, a direct comparison between radial distance measurements d of the lidar sensor and expectations on the template $\widetilde{d}$ is possible. The mean squared error E for the translations t_x and t_z quantifies their deviation

$$E_{t_x,t_z} = \frac{1}{|\mathscr{I}|} \sum_{i \in \mathscr{I}} (\widetilde{d}_i - d_i)^2 \quad \text{with} \quad \mathscr{I} = f_{\text{shape}}(t_x, t_z), \tag{2.16}$$

where only measurements i that are located on the rail shape are considered. Finally, a binary classifier decides whether the measurements resemble the template as proposed in [Hackel et al. 2015]. It evaluates the minimal deviation of all translations within the search space

$$\min_{(t_x,t_z) \in \mathscr{S}_k} (E_{t_x,t_z}) < \lambda_{\text{shape}} \tag{2.17}$$

and detects a rail profile if this deviation is below a threshold λ_{shape}. Exemplary results of the template matching are shown in Figure 2.14.

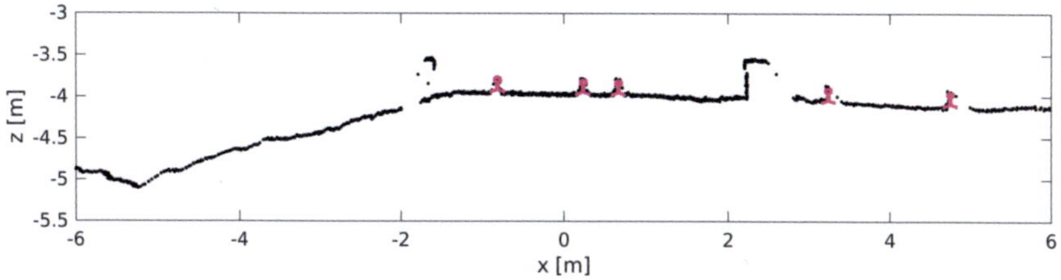

(a) Five correctly detected rail profiles (magenta; cp. also Figure 2.4b).

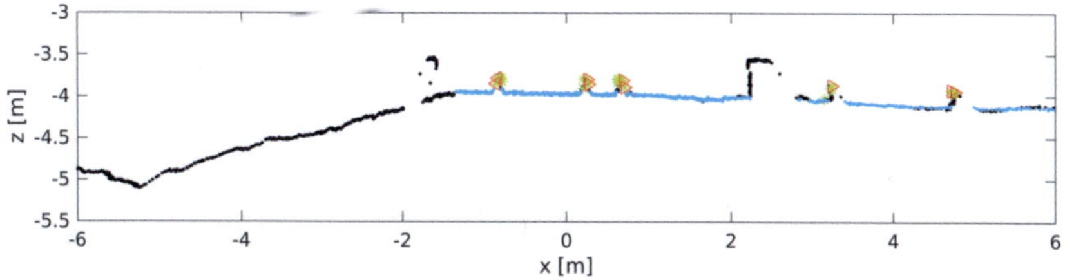

(b) Used features which have been derived in previous steps: ground plane (blue dots), occluding edges (red triangles), and echo key points (green asterisks).

Figure 2.14: Exemplary rail profile detection results. Black dots represent measurements.

Remarks: Since the shape of the rail is a priori known, the template library can be initially determined by simulation, using for example a ray casting approach. For an efficient implementation, the template library only contains measurements that are located on the template (cp. f_{shape} in (2.15) and (2.16)). Moreover, the detection area is subdivided equidistantly in x- and z-direction with resolutions Δt_x and Δt_z respectively where $\Delta t_x = \Delta t_z = 1\,\text{cm}$ is sufficient for this work.

Note that [Blug et al. 2004] require that each object has at least to be hit by three measurements. In contrast, this approach covers — depending on the translations t_x and t_z — from 27 to 42 measurements that are located on an UIC 60 rail profile within the proposed detection area, while assuming an angular resolution of $\Delta\varphi = {}^1/_{14}{}^\circ$. It significantly exceeds this minimal requirement and is thus robust against a few missing measurements or outliers. By considering all feasible translations within the search space, the approach does not get stuck in a local minimum, but guarantees a globally optimal solution.

2.4 Detection of rails and tracks within consecutive scans

After the shapes of rails have been detected within vertical cross-sections of the nearby environment (x-z-plane), their appearance in longitudinal dimension (y-direction) is considered. Individual rail profile detections are observed over consecutive scans and spatially clustered to rail sections at first. Then, tuples of parallel rail sections are validated for plausibility and arranged in such a way that track sections can be derived[21]. This corresponds to the second and third step of the multistage approach proposed in Figure 2.2.

2.4.1 Detection of rail sections

Idea: Rails are elongated objects which can be straight or curved (cp. Figure 2.1). Their common features are identified at first. Straight rails are linear by definition. They hardly change their position relative to the lidar sensor between consecutive scans as long as they are parallel to the driving direction of the railway vehicle. If they are diagonal to the ego track, they have a linear course in the x-y-plane (i.e., in a top view). Even curved tracks cannot change their course abruptly, since they need to be continuous for design and comfort reasons. Thus, rails can be assumed to be approximately linear within short periods when they are scanned by the proposed setup[22]. A method is required that decides whether detections of rail profiles within consecutive scans belong to the same rail or not. More than one rail section is typically located within a scan.

[21] The term "section" refers to the fact that only parts of such elongated objects can typically be observed. Their starts and ends do not necessarily correspond to their physical ones, since they might be partially occluded or not observed within the considered detection area.

[22] Even when the railway vehicle drives through a curve with the maximal allowable velocity, the lateral deviation of the curved track compared to a straight track is less than 1 m within 1 s. However, within this period of time, there are at least 50 scans which perceive the nearby environment within vertical cross-sections. Thus, the lateral deviation between consecutive scans is just a few centimeters, while a rail has a width of more than 10 cm (cp. Table 2.1). Moreover, the dynamics of railway vehicles are limited. The maximal acceleration is typically much smaller than $1\,{}^{m}/_{s^2}$. Thus, the velocity can be assumed as constant within such a period.

Method: To group rail profile detections that have been detected in several scans a spatial clustering approach is used as proposed in [Hackel et al. 2015; Stein et al. 2016b]. It is based on seeded region growing which is a fast bottom-up approach that identifies homogeneous clusters [Adams and Bischof 1994]. It allows to incorporate model knowledge, such as the piecewise linearity of rail courses. Its basic structure, the treatment of ambiguity, and the used similarity measure are considered subsequently

Algorithm 2.1: Basic structure of the spatial clustering algorithm (sliding window size ω, maximal deviation δ, minimal number of observations ν, number of scans O; based on [Stein et al. 2016b]).

procedure RAILSECTIONDETECTION(ω, δ, ν)
  $C_{\text{all}} \leftarrow \emptyset$ ▷ set of all clusters
  $C_{\text{window}} \leftarrow \emptyset$ ▷ clusters in sliding window
  for scans $s = n^{(1)} \ldots n^{(O)}$ **do**
    for all rail profile detections r in scan s **do**
      $(c_{\min}, d_{\min}) \leftarrow$ CLOSESTCLUSTER(r, C_{window})
      if $d_{\min} \leq \delta$ **then**
        $C_{\text{all}} \leftarrow C_{\text{all}} \setminus \{c_{\min}\}$
        $c_{\min} \leftarrow c_{\min} \cup \{r\}$ ▷ add to existing cluster
      else
        $c_{\min} \leftarrow \{r\}$ ▷ create new cluster
      end if
      $C_{\text{all}} \leftarrow C_{\text{all}} \cup \{c_{\min}\}$
    end for
    $C_{\text{window}} \leftarrow \{c \in$ RECENTCLUSTERS$(s, C_{\text{all}}, \omega) \mid |c| \geq \nu \vee \ldots$
        $s - \omega + 1 <$ FIRSTOBSERVATION$(c) \leq s\}$
        ▷ ignore old detections and tiny clusters
  end for
  return C_{all}
end procedure

Algorithm 2.1 shows the basic structure. Every rail profile detection of the first scan is an initial seed. Thus, it is a cluster, which describes a potential rail section. Then, the approach works incrementally, is only applied in driving direction, and is thus causal. Based on the last scans, a decision is made for every further rail profile detection in every consecutive scan. If a detection fits to a current rail section, it is added. In this way the region grows. Otherwise, a new cluster has to be established. To avoid endless sequences and to correspond to the assumption of piecewise linear courses, only rail profile detections within a sliding window of at most ω scans are considered (cp. RECENTCLUSTERS in Algorithm 2.1). Furthermore, current clusters need to have a minimal number of observations ν within this sliding window or they are not further considered. Recently created clusters remain (cp. FIRSTOBSERVATION in Algorithm 2.1). Rail profile detections are added to the nearest existing cluster as long as they are not more than a maximal deviation δ away from it. Moreover, every cluster is extended at most by one rail profile detection per scan[23], since rails are assumed to be linear (cp. Figure 2.15). However, in the case of a splitting from one rail into two rails, which occurs for example on turnouts, the recent cluster follows the closest detection. The other detection establishes a new cluster.

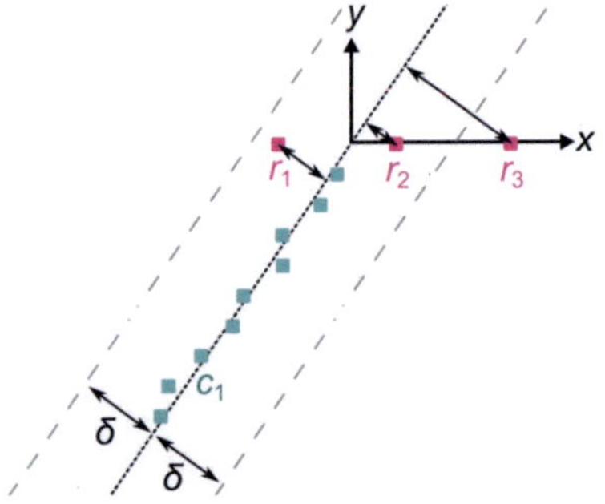

Figure 2.15: Idea of seeded region growing (top view). Each of the 12 squares represents a rail detection: three magenta ones (denoted as r_i) are detected in the current scan, turquoise ones have already been assigned to the only cluster c_1 (main direction represented by dotted line) within the sliding window of $\omega = 10$ scans. Detection r_3 is too far away from c_1, thus it is represented by a new cluster. In contrast, detections r_1 and r_2 are less than the maximal deviation δ away from c_1. However, only the nearer r_2 is added, while another cluster is created for r_1.

[23] Note that this is not explicitly shown in the basic structure of Algorithm 2.1.

For a clear allocation to a cluster, a similarity measure is needed. The corresponding method CLOSESTCLUSTER returns the nearest cluster and the smallest distance from it. The orthogonal distance[24] from the main direction of the closest cluster is used. A principal component analysis (PCA) determines the main direction of previous clusters as proposed in [Yang and Fang 2014]. Exemplary rail section detection results are shown in Figure 2.16.

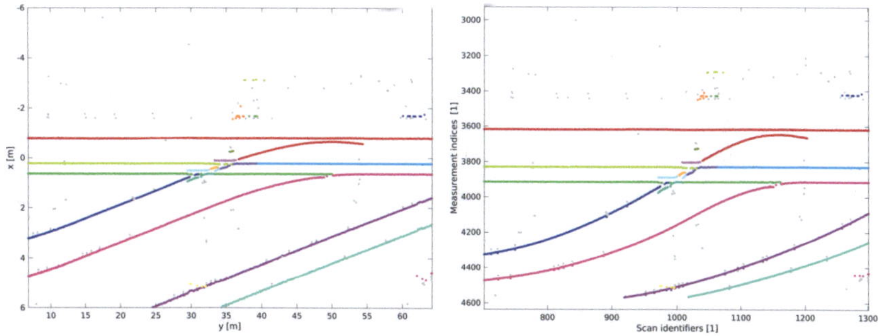

(a) Metric representation.

(b) Index representation.

Figure 2.16: Exemplary rail section detection results. The rail profile detections (dots represent their centers) from Figure 2.14a are shown on the left side (as the scan with minimal y-value or minimal scan identifier respectively). Both the Cartesian representation in the (metric) x-y-plane and the sensor representation in indexes of measurements and scans is shown (cp. also Section 2.2.2). The movement of the railway vehicle in y-direction, which is for example determined by an additional odometer, is only shown for illustration purposes, but not considered in the proposed approach. Valid clusters have an individual highly saturated color, but the color might be used repeatedly. Too small clusters (gray) are not further considered. Note that especially wing rails in the center are assigned to individual clusters, since their orientation differs from the other rails. A few small clusters occur outside of the tracks.

Remarks: The minimal requirement of at least three measurements per object from [Blug et al. 2004] applies for the longitudinal direction of rails as well. However, to avoid tiny, but erroneous clusters, $\nu = \omega/10$ is required herein where a sliding window size of 1 s is a good choice (i.e., $\omega = 50$ or more scans, depending on the measurement rate f). Even a segmentation of one rail into

[24] The Euclidean distance is used instead, if a cluster only contains one detection within the sliding window.

several clusters is uncritical and can be taken into account in subsequent steps. In contrast, an interpolation or extrapolation is not expedient, since rails can actually be split (cp. the missing parts of a straight track in a real world scenario on the right side of Figure 2.6). Thus, short periods of missing detections are taken into account, while a new cluster is created otherwise. Whereas [Hackel et al. 2015] required 3d information to evaluate similarity, distances in the *x*-*y*-plane are replaced by the knowledge of the temporal and spatial arrangement of the measurements[25].

2.4.2 Detection of track sections

Idea: A track is an elongated object. It is formed by parallel rails (typically a pair of rails) that have a certain lateral distance (cp. gauge G in Figure 2.17). Otherwise, the guided railway vehicle might derail. Thus, a method is required that decides whether and which previously detected rail sections (i.e., clusters) belong to the same track section as proposed in the overall structure in Figure 2.2. At least one track section that belongs to the ego track is expected in a scan.

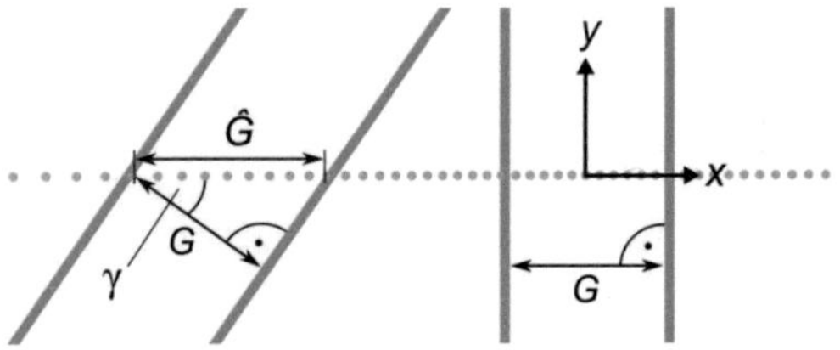

Figure 2.17: Exemplary gauge observation of two tracks (top view). Since the railway vehicle runs on the right track, the left track is observed diagonally with oblique angle γ. Thus, an increased value $\widehat{G}$ of the actual gauge G is observed in a single scan (orange).

[25] The corresponding equidistant grid resembles the chronological sequence of measurements within a scan (index i, cp. (2.1)) and between scans (identifier $n^{(j)}$, cp. (2.3)) as shown in Figure 2.16b. Linearity can be examined heuristically in this way. Finally, the mean index of all measurements on the rail profile is used as reference of every rail profile detection.

Method: Based on the geometrical and topological characteristics of a track, an incremental detection approach is proposed, which consists of two substeps and only applies in driving direction. It decides upon the potential side of a rail in the track (i.e., left or right) at first. Consecutively, it determines whether and which rail sections form a track section. In addition, it is shown how potential ambiguities are treated. First of all, the lateral search area is determined, which is used to find parallel counterparts of the previously detected rail sections.

Since this approach aims to detect track sections immediately, a decision after each scan is required as previously done for rail sections. However, the gauge cannot be measured by definition (cp. Section 2.3.1), since one side of the rail might be occluded[26]. Moreover, tracks besides the ego track might be observed diagonally from the lidar sensor. In that case, the distance between both rails in a scan does not equal the gauge G, but has a larger value $\widehat{G}$ depending on the oblique angle γ between the lidar sensor measurements and the track (cp. the left track in Figure 2.17):

$$\widehat{G} = \frac{G}{\cos(\gamma)}. \tag{2.18}$$

As a result, the gauge is only approximated within one scan. To avoid that every cluster is a potential partner of every other cluster (i.e., $\widehat{G} \to \infty$ for $|\gamma| \to 90°$), the absolute value of the considered oblique angle is further limited by $\gamma_{\max}$. Thus, the lateral search area for corresponding rail sections ranges from the minimal gauge (e.g., $G_{\min} = 1430\,\text{mm}$ for normal gauge) to the maximal considered, but diagonally measured gauge (e.g., $G_{\max} = 1470\,\text{mm}$ observed under oblique angle $\gamma_{\max}$ in (2.18)).

[26] Note that only 2d information from the lidar sensor in the x-z-plane is considered herein. However, if distances between consecutive measurements in y-direction are available, parallelism can be checked additionally as done for example in [Hackel et al. 2015].

Algorithm 2.2: Basic structure of the detection of track sections (detected rail sections C_{all} from Algorithm 2.1, sliding window size ω, minimal number of observations ν, minimal and maximal gauge $G_{\text{min}}, G_{\text{max}}$, maximal absolute oblique angle γ_{max}, number of scans O).

```
procedure TRACKSECTIONDETECTION(C_all, ω, ν, G_min, G_max, γ_max)
    S_window ← ∅                              ▷ potential cluster sides in sliding window
    T_all ← ∅                                                     ▷ all detected tracks
    for scans s = n^(1) ... n^(O) do
        C_window ← {c ∈ RECENTCLUSTERS(s, C_all, ω) | |c| ≥ ν ∨ ...
                    s − ω < FIRSTOBSERVATION(c) ≤ s}
          ▷ only consider new or sufficiently large clusters, cp. Algorithm 2.1
        for all clusters c in scan s do
            S_window ← POTENTIALRAILSIDE(s, c, S_window, C_window, ω, ...
                                          G_min, G_max, γ_max)
            if s < n^(ω) then
                continue                      ▷ initialization phase of ω − 1 scans
            end if
            (S_window, T_all) ← RAILASSOCIATION(s, c, S_window, C_window, ...
                                          T_all, ω, ν, G_min, G_max, γ_max)
        end for
    end for
    return T_all
end procedure
```

In the first detection substep, the potential side of a rail in a track is determined based on the lateral arrangement of the clusters. Besides “left” and “right” rails in a track, possible decisions also include “undecided” and “irrelevant”. A cluster is irrelevant if no corresponding cluster to its left or right side can be found within the sliding window of ω scans. In contrast, a cluster has an undecided side as long as it is not considered in a track section or if the current arrangement contradicts the previous side. Since the shape of Vignole rails is symmetric, it cannot directly be decided on this side from their lidar sensor measurements. For example, the second rail from the left in Figure 2.17 has a similar lateral distance from the rail on its left and on its right side. This means that it might fit to both rails if only the detections within the current scan were evaluated. Thus, the previous side of each rail section is also considered (cp.

POTENTIALRAILSIDE in Algorithm 2.2). However, the final decision is made when a track section is detected. Again, the decision only bases on the detections in the sliding window of ω scans (cp. the rectangles in Figure 2.18a, that exemplarily represent the size of the lateral search area and the (longitudinal) sliding window for one rail) where the first $\omega - 1$ scans are used to initially determine potential sides of rails in a track (cp. Figure 2.18b).

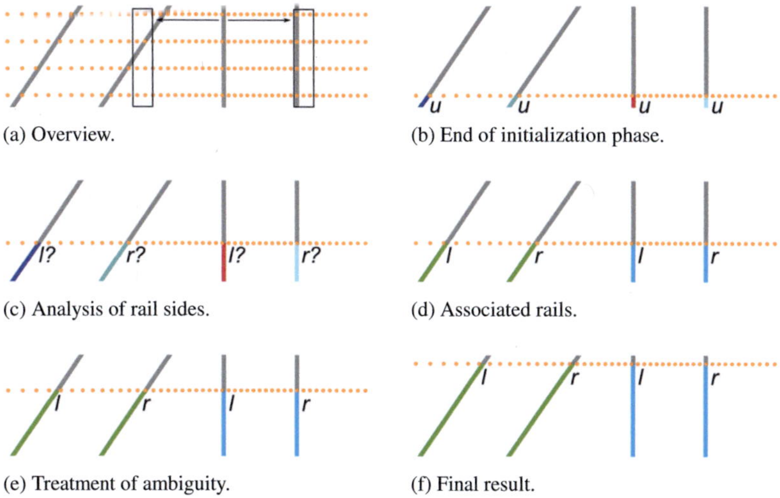

(a) Overview.
(b) End of initialization phase.
(c) Analysis of rail sides.
(d) Associated rails.
(e) Treatment of ambiguity.
(f) Final result.

Figure 2.18: Idea of track section detection (top view; selected measurements in orange, rails in gray, potential rail sides l (left), r (right), and u (undecided)). Each rail and track section has their individual color. Both rectangles in (a) exemplarily represent lateral and longitudinal search areas of the second rail from the right for potentially associated rails on the left and the right side assuming a sliding window of four scans. It can be clearly seen, that the rail to its left is only once within the lateral search area. Thus, the blue track section is continued in (e) and (f).

The second detection substep decides for every rail section directly after each scan whether it belongs to a track section or not and on which side the rail section is therein (cp. RAILASSOCIATION in Algorithm 2.2, Figure 2.18c, and Figure 2.18d). If a combination of rail sections fits to an existing track section, it is continued, otherwise a new track section is created. Similar to rail sections, at most one detection is added to a track section per scan. In contrast, track

sections might split on turnouts. In order to avoid erroneous decisions, rail sections are only considered when they at least have ν observations within the sliding window ω. In this way track sections grow, new ones arise, and no longer relevant ones terminate.
Every potential combination to the left and/or right in the lateral search area is evaluated for every cluster. Ambiguous combinations need to be treated where rails might be used either on the left or the right side of a track or are a potential partner for several tracks (cp. again the second rail from the left in Figure 2.17, which has a similar lateral distance from the rails to its left and right side). Thus, rail sections are only assigned to a track section if a clear decision of both involved rail sections is made (cp. Figure 2.18e and Figure 2.18f). Exemplary results of the detection of track sections are shown in Figure 2.19.

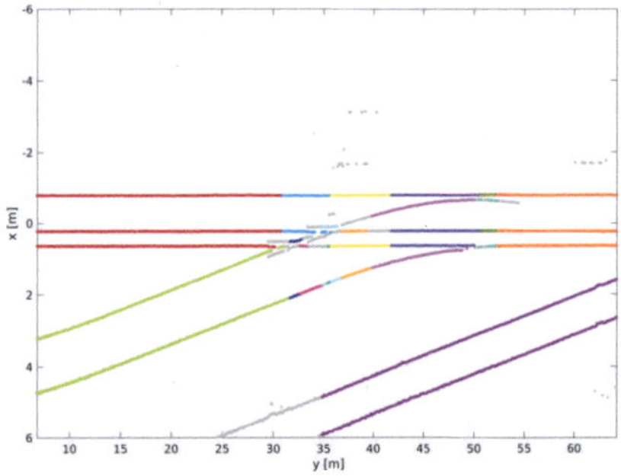

Figure 2.19: Exemplary track section detection results. Detected track sections have an individual highly saturated color where the colors are completely independent from those of the rail sections in Figure 2.16. Dots represent the centers of the corresponding rail profile detections. Unused parts of rail sections (gray) are not further considered. Note that the assignment of rail sections to track sections typically changes only on turnouts due to the variety of potential combinations. Clusters which are detected outside of the tracks have not been used at all herein.

Remarks: The minimal number of measurements per object introduced by [Blug et al. 2004] applies again. As for the rail section detection, a minimum of $\nu = \omega/10$ observations is required before a track section is valid, whereby the sliding window size equals 1 s again (i.e., $\omega = 50$ or more scans, depending on the measurement rate f, whereby significant changes of a track are not expected therein).

Since the gauge cannot be determined by definition within a scan, the lateral distance between the axes of symmetry of both rail profiles is used. Their values t_x are directly known from template matching.

The maximal absolute value of the oblique angle should be chosen so that $\widehat{G}$ is always smaller than the minimal lateral distance of parallel tracks $x_{\mathrm{t,min}}$ (e.g., $x_{\mathrm{t,min}} = 3.5\,\mathrm{m}$ for railway tracks in Germany that have normal gauge $G = 1.435\,\mathrm{m}$; cp. also Figure 2.5a), otherwise further ambiguities would occur. $\gamma_{\mathrm{max}} = 39°$ is a good compromise (as shown below).

The decision on a side or on the belonging to a track section is implemented as a majority vote, whereby recent observations have a stronger weight. Every potential cluster within the lateral search area counts where several votes are possible per considered detection and scan depending on the overall arrangement. Since parallelism, fixed lateral distance, and similar direction cannot be evaluated in a scan, they are checked by the similarity of the observed lateral distances $\widehat{G}$ between consecutive scans within the sliding window.

Extension option for various gauges: In rare cases, multiple rails form a track. For example, a mixed gauge track with three rails has two rails on the left (or right) side (cp. Figure 2.20b and Figure 2.20c respectively). Their lateral distances to the right rail (or left rail) are those of two gauges. Thus, it allows for example the running of both normal and meter gauge railway vehicles (cp. Figure 2.20a and Figure 2.20d).

An extension from two to three rails per track is considered herein (cp. for example also the red track section in Figure 2.19). However, the proposed approach can be adapted analogously for every rail arrangement. Only RAILASSOCIATION needs to be extended for modified rail arrangements within a track. It is the first step of the overall proposed approach that is influenced by various gauges. It examines all combinations shown in Figure 2.20 for their detection and checks whether the scaling of both observed gauges is similar. In addition, and identical to common tracks, their development between consecutive scans is evaluated.

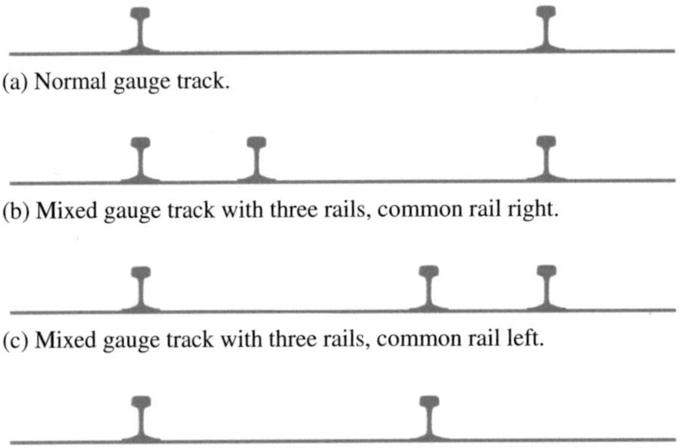

(a) Normal gauge track.

(b) Mixed gauge track with three rails, common rail right.

(c) Mixed gauge track with three rails, common rail left.

(d) Meter gauge track.

Figure 2.20: Exemplary track types (cross-sections). Rails for meter gauge ($G = 1000\,\text{mm}$) and/or normal gauge ($G = 1435\,\text{mm}$) vehicles are shown.

Finally, the maximal oblique angle γ_{max} allows to distinguish between both gauges, whereby its previously proposed value of 39° is the maximal value[27]. In this way a track can be classified additionally whether it has meter or normal gauge or is a mixed gauge track with three rails (and whether the latter one has its common rail on the left or the right side).

2.5 Inference from detected rails and tracks

To conclude on the topology of the railway network and the movement of the railway vehicle therein, the previously detected rail and track sections are further processed. Firstly, the lateral arrangement of several tracks is evaluated.

[27] In contrast, if γ_{max} is increased by another degree, the maximal observed meter gauge exceeds the minimal normal gauge of 1430 mm (maximal meter gauge (1035 mm) plus twice the lateral offset of the symmetry axis of the templates (equals in total the width of the head of the rail; 172 mm) observed under oblique angle γ results in the observed gauge following (2.18)).

Secondly, positions on the turnout are determined and the branching direction thereon is detected. This corresponds to the fourth and fifth step of the proposed multistage approach (cp. its overall structure in Figure 2.2) and completes the information which can solely be obtained from lidar sensor measurements. Thereby derived and further provided information will be called event, which is motivated by [Rahmig et al. 2013].

2.5.1 Detection of topology events

Idea: Based on the previous steps of the multistage approach, it is already known whether and which rail section belongs to the same track section and how they are arranged therein (i.e., whether they are a left or a right rail). Several tracks can be arranged in different ways to each other, such as parallel, diagonal, or intersecting, and at varying lateral distances (cp. Figure 2.1). Their lateral arrangement provides information on the topology within the nearby environment of the railway vehicle.

Method: The proposed approach detects the ego track and classifies all other tracks into those on the left and on the right side. In addition, a compact representation can be created, which consists of topology events (cp. Algorithm 2.3). Then, the latter ones can be fed into an application, such as a train-borne localization system. In every case, a decision is again made directly after each scan. To avoid premature and erroneous decisions, only (track) sections which at least have ν observations within the sliding window ω are further considered as already proposed in Algorithm 2.1 and Algorithm 2.2 (cp. RECENTTRACKS and FIRSTOBSERVATION in Algorithm 2.3). Based on this, the ego track can be determined (cp. EGOTRACK). Initially, this is the track in the current scan that is closest to the center of the x-axis. Previous decisions on the ego track are considered in subsequent scans, whereby the track with the minimal lateral deviation from the center of the previous ego track is chosen.

Based on the knowledge of the ego track and its lateral center, lateral distances from every track can be determined (cp. NEARBYTRACKS in Algorithm 2.3).

Algorithm 2.3: Basic structure of the inference of topology events (detected track sections T_{all} from Algorithm 2.2, sliding window size ω, minimal number of observations ν, maximal gauge G_{max}, number of scans O).

```
procedure TopologyDetection(T_all, ω, ν, G_max)
    x̄_ego,0 ← 0.0                                    ▷ default center of ego track
    E_all ← ∅                                                    ▷ all ego tracks
    L_all ← ∅                                                   ▷ all left tracks
    R_all ← ∅                                                  ▷ all right tracks
    N_all ← ∅                                               ▷ all topology events
    for scans s = n^(1) ... n^(O) do
        T_window ← {t ∈ RecentTracks(s, T_all, ω) | |t| ≥ ν ∨ ...
                    s − ω < FirstObservation(t) ≤ s}
                      ▷ only consider new or sufficiently large track sections
        E_all = EgoTrack(s, x̄_ego,0, E_all, T_window, G_max)
        for all track sections t in scan s do
            (L_all, R_all) ← NearbyTracks(s, t, E_all, L_all, R_all, T_window)
            N_all ← TopologyEvents(s, t, E_all, L_all, R_all, N_all, T_window, ω, ...
                                   ν, G_max)
        end for
    end for
    return (E_all, L_all, R_all, N_all)
end procedure
```

Current tracks, whose x-value of the track center is smaller than that of the ego track, are classified as left tracks[28]. All tracks with a center whose x-value exceeds those of the ego track are located on the right side. A side can be determined for every observation of a track. However, this assignment can change over time, for example from left to right if this track intersects the ego track on a diamond crossing (cp. also Figure 2.1b).

The information on the nearby environment can be further compressed (cp. TopologyEvents in Algorithm 2.3). In order to avoid endless or erroneous

[28] This side is defined as observed by the lidar sensor. If the railway vehicle drives backwards, this track is located on the opposite side in driving direction. Such a reversing cannot be detected in lidar sensor data. Nonetheless, the overlying application usually has information on the movement and can thus adapt the sides where appropriate.

topology events, each left or right track is only reported once within the sliding window of ω scans. Furthermore, at least v observations therein are required for such an event. Relevant characteristics of the corresponding topology event are summarized in Table 2.2. Its mandatory part includes the point in time[29] of the first observation, which is relevant for this event, its detection time, and the detected side. Additionally, the centers of the ego and the reported track can be included, which allow to determine their lateral distance. Finally, to characterize the relevance of an event, the number of observations since the last event and for all the time can be integrated. Figure 2.21 shows exemplary topology detection results.

Table 2.2: Contents of a topology event. Optional entries are *italic*.

category	**entry**	**values**	**unit**
points in time for **this** event	first observation	double	s
	detection	double	s
arrangement	side	left | right	none
	center ego track	*double*	*m*
	center this track	*double*	*m*
number of observations	*this event*	*unsigned integer*	*none*
	overall	*unsigned integer*	*none*

[29] Points in time are used instead of distances in y-direction, since the latter ones cannot be determined from the considered 2d lidar sensor data. However, the overlying application can synchronize those points in time with corresponding motion and position measurements and thus determine the (global) location of the reported detection. Assuming a constant measurement rate f, even the typically ascending identifiers from (2.3) can be used. In this way identifiers can be converted into relative time stamps.

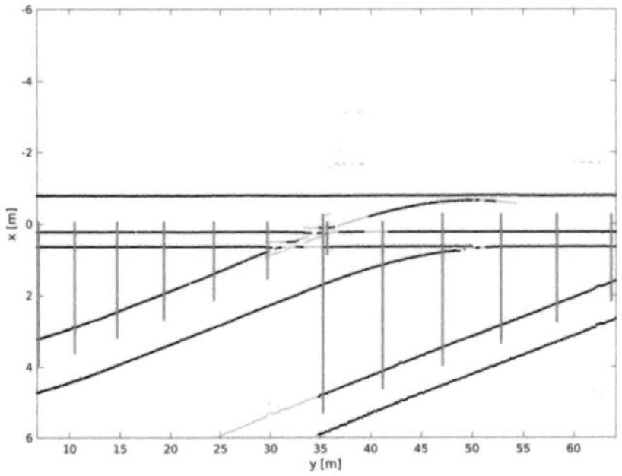

Figure 2.21: Exemplary topology detection results. Vertical green lines indicate the position in y-direction where a topology event on the right side of the ego track has been detected, while their start and end point in x-direction represent the estimated track centers of the ego and the other track. Black dots represent rail centers of detected track sections from Figure 2.19, whereas gray dots correspond to centers of excluded rail sections. Although the assignment of rail sections changed on the turnout, topology events have been detected accurately, while their distance in y-direction rises with increasing velocity of the railway vehicle.

Remarks: The parameters ω, ν, and G_{max} are chosen as before. Since the lateral center of a track cannot be determined by definition in a single scan, both symmetry axes of the corresponding rails are used again. In this way the center point and thus lateral distances $\widehat{x_t}$ between tracks are approximated (cp. Figure 2.22). Since turnouts are detected in the subsequent step, tracks which most probably have at least one rail within the ego track (i.e., their center is located within $|x| \leq G_{\text{max}}/2$) are excluded from topology events. This comprises the ego track and nearby other tracks, whereby the latter ones are that close only within short sections. Thus, this exclusion causes no further limitations, while the information on turnouts is reported by subsequent branching events.
A smaller fraction of the maximal gauge G_{max} can be used to restrict the lateral search area for the ego track, which especially results from deviations of the lidar sensor relative to the track center. The optional number of observations in Table 2.2 is an indicator for the certainty of the current event, which has been proposed, but not further specified in [Rahmig et al. 2013]. Finally, to achieve maximal (stochastic) independence of topology events, track sections are not considered for reporting as long as any of their rail sections (i.e., the

smallest, not independent detection) is already used for a topology event within the sliding window.

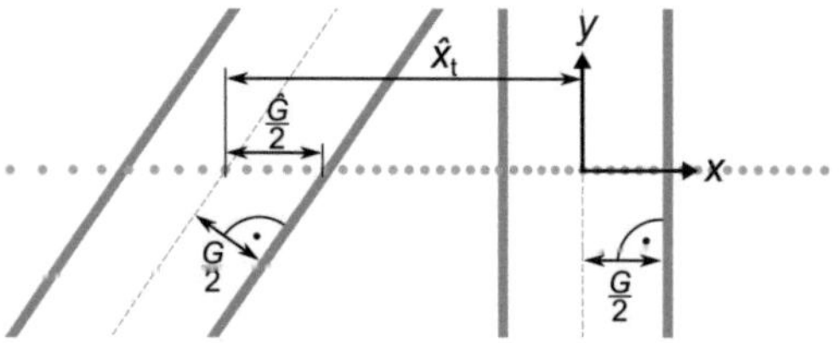

Figure 2.22: Exemplary lateral distance observation between two tracks (top view). The railway vehicle runs still on the right track and the left track is thus observed diagonally in a single scan (measurements in orange). The lateral track distance $\widehat{x}_t$ can be determined from the possibly increased values $\widehat{G}$ of the gauges G.

Extension option for various gauges: The proposed topology event in Table 2.2 can even be extended to the additionally considered mixed gauge tracks with three rails (cp. the (horizontal) ego track in Figure 2.21). Since several kinds of tracks exists (cp. the four ones shown in Figure 2.20), their detected type can be reported additionally. Moreover, the center of the track can be determined for meter and/or normal gauge (i.e., a mixed gauge track with three rails has for example a center for meter gauge and a center for normal gauge, while a common normal gauge track only has a center for normal gauge and none for meter gauge). Those extensions for the track type and different center points apply for both the ego track and the reported nearby track[30], whereas further adaptations are not required for Algorithm 2.3.

[30] Moreover, depending on the type of the track, lateral distances between tracks vary. A mixed gauge track with three rails and the common rail on the left side has a larger distance from a detected meter gauge track on the right side compared to a mixed gauge track with three rails and the common rail on the right side. However, if tracks have several gauges, the greatest common gauge should be used for determining $\widehat{x}_t$ from Figure 2.22.

2.5.2 Detection of the branching direction on turnouts

Turnouts lead to ambiguity problems, which have been described in Chapter 1. However, the knowledge of their position would help to adjust the estimated position of the railway vehicle. In order to detect turnouts on the ego track and the branching direction thereon, the most common turnout type, i.e., single turnouts, is considered herein. Moreover, those turnouts should be preferred to other designs due to their simplicity [Matthews 2007]. Several types of construction exist (cp. Figure 2.23), while all have in common that they allow branching from one track into two tracks. Therefore, in the following, each of them is referred to as a single turnout, so that it is independent of whether one (cp. Figure 2.23a and Figure 2.23b) or both (cp. Figure 2.23c and Figure 2.23d) of the branching tracks are curved. Note that the curvature of the branching tracks affects the maximal allowed velocity, i.e., the more elongated the turnout is, the faster a railway vehicle can pass the turnout on this curved track.

(a) Straight and curved track, left-hand.

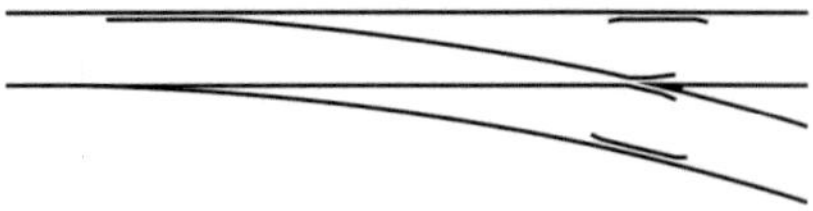

(b) Straight and curved track, right-hand.

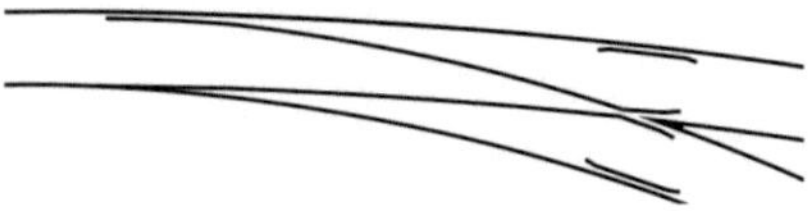

(c) Both tracks curved in same direction.

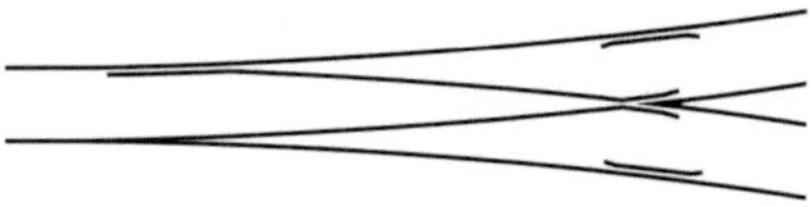

(d) Both tracks curved, but with different directions (wye switch).

Figure 2.23: Types of single turnouts.

Two reference positions on single turnouts are considered at first. Finally, it is detected in which direction the railway vehicle passed the turnout and how corresponding events can be created from it.

Detection of the reference position on turnouts

Idea: Typical elements of a single turnout are switch blades, check rails, wing rails, and the frog (cp. Figure 2.24). They occur independent of whether one or both tracks are curved and whether the single turnout is left-hand or right-hand. The frog and the switch blades are considered next, since they are distinctive points for a detection of the turnout. In contrast, the railway vehicle does neither drive on wing rails nor on check rails, but they prevent derailing.

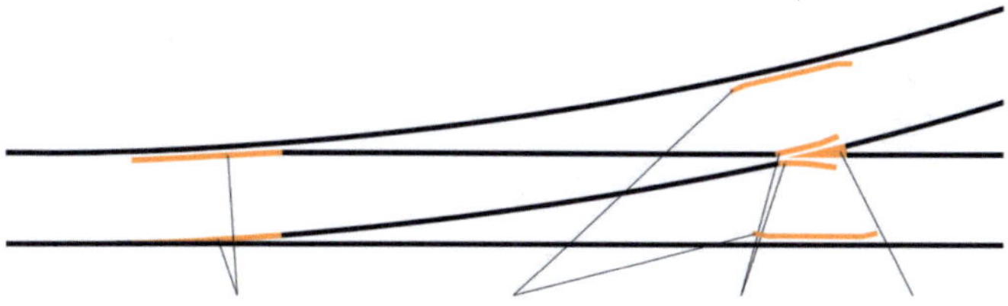

Figure 2.24: Relevant elements of a single turnout. Switch blades are movable.

When driving in reading direction in Figure 2.25, the lidar sensor observes a blade rail at first and later on a check rail on the right side as well as the wing rail and the frog on the left side. The tip of the frog is the first point where the gauges G of the ego track and the branching track do not overlap (cp. the turquoise and gray areas in Figure 2.25a). Moreover, the railway vehicle drives by the frog each time when it passes a turnout. This is independent of the direction taken thereon.

The switch blades are movable and thus determine the branching direction. However, only one side of them is striking, since it is located within the ego track (cp. the orange part in Figure 2.25b), while the opposite side is part of the ego track in this scenario. Following the idea of [Hackel et al. 2015], this orange part of the switch blades is characterized as the rail, which is located within the ego track and has a lateral distance of less than the minimal gauge $G_{\min}$ from it (cp. also the yellow area in Figure 2.25a). Furthermore, its distance Δx_{b} from the nearby rail of the ego track at least is 100 mm for design reasons [Fiedler

1999]. Thus, this blade rail differentiates from other rails. The position of the tips of frog and blade are detected subsequently.

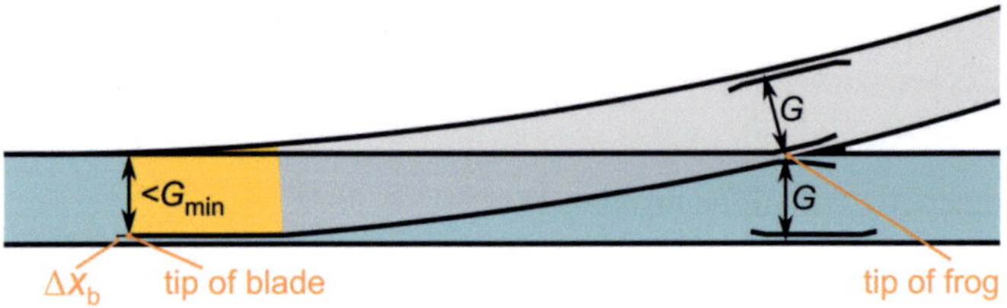

(a) Reference positions.

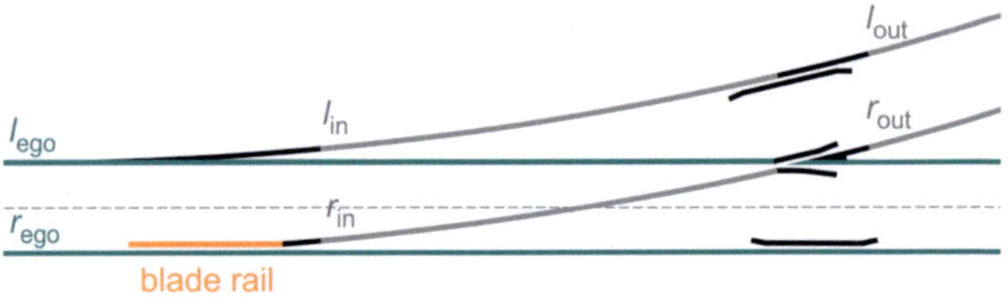

(b) Detections on turnouts.

Figure 2.25: Idea of single turnout detection. In (a) turquoise and gray areas represent sections with a valid gauge G. In contrast, the yellow area has a width of less than G_{min} for design reasons, since the railway vehicle could otherwise not pass the turnout on the gap denoted with Δx_{b}. In (b) relevant detected track sections are shown: ego track (turquoise) and branching track (gray), which is at first partially inside and then completely outside. Left rails are denoted by l and right rails by r.

Method: The frog provides the basis for the detection of a turnout. In order to determine its tip, already detected track sections from Algorithm 2.2 can be used. Moreover, their assignment to ego, left, and right tracks from Algorithm 2.3 allows to distinguish their arrangement. The relevant track sections for the turnout detection include the ego track as well as the branching track (cp. the turquoise and gray parts in Figure 2.25b respectively). The branching track has a part which is partially located inside the ego track and another part that is completely outside of the ego track. The intersection point of the rail extensions which are closest to the frog determines the position of its tip. For the example shown in Figure 2.25b, this includes the right rails of the branching track (denoted by r_{in} and r_{out}) and the left rail of the ego track (denoted by l_{ego}). In order to avoid premature decisions, at least v observations are required for both the inside and the outside part of the branching track. Thus, the detection of the tip of the frog always occurs after its passing. Exemplary detection results are shown in Figure 2.26a.

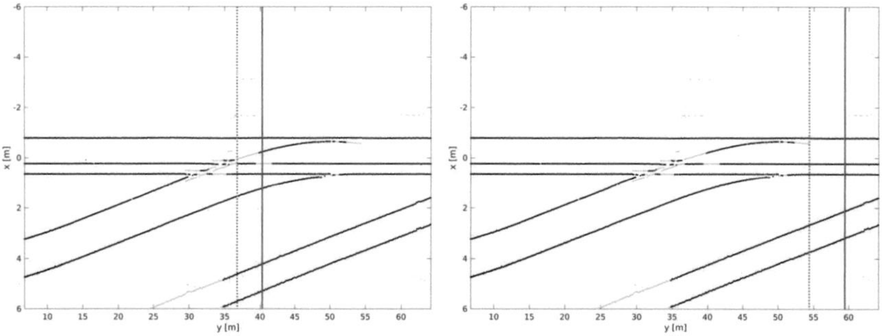

(a) Detection of the tip of the frog.

(b) Detection of the tip of the blade.

Figure 2.26: Exemplary reference position detection results on single turnouts. Vertical solid lines indicate the position where the reference position has been detected, while dotted lines represent the position of the estimated tip. The color of each line indicates the detected branching direction where purple represents trailing right. Black dots represent rail centers of detected track sections from Figure 2.19, whereas gray dots correspond to centers of excluded rail sections. While the estimated frog position is imprecise, the tip of the blade fits its actual position, even though the tip is not part of the detected branching track and thus shown in gray.

However, the position of the frog only has minor significance for the branching detection. For this, the switch blades are the relevant part on turnouts, since their setting determines the route taken on a turnout. This means that the detection of the frog is a necessary prerequisite. In Figure 2.25a, the tip of the blade is the first observed point on the right rail of the branching track. In order to avoid premature or erroneous decisions, this point is only classified as tip of the blade when at least v detections on r_{in} are observed on both sides of the center of the ego track (besides the v observations on the part of the branching track that is completely outside of the ego track as previously required for the frog detection). In this way potential blades can be distinguished from other rails within the ego track, such as check rails. The latter ones are either located left or right of the center of the ego track. Moreover, the detection of these reference positions on single turnouts provides the basis for the detection of the branching direction thereon. Exemplary blade detection results are shown in Figure 2.26b.

Remarks: This approach applies for different types of single turnouts analogously. It is irrespective of whether the single turnout is right-hand instead of left-hand, if both tracks are curved, or if it is passed from the opposite side where the sides of the considered rails might change. Since the longitudinal dimension of the single turnout varies and this step is no longer focusing on the linearity of objects, the minimal number of observations ν is not limited to the sliding window of ω scans.

Although the frog is always passed on single turnouts, its position can only be approximated herein[31]. This results from the detected parts of the branching track that do typically not include the tip of the frog itself. There might be no detection of the branching track near the tip of the blade, since the corresponding Algorithm 2.2 considers the minimal gauge G_{min}. Thus, the beginning of the blade rail which has been detected in Algorithm 2.1 can be used instead of the beginning of the branching track[32].

Extension option for various gauges: This approach can even be used for complex turnouts which also have mixed gauge tracks with three rails. Since several frogs might occur for design reasons, the outermost frog, which results from the intersection of the largest gauges of each connected track, can be detected similarly to the proposed approach. The same applies to the tip of the blade where that with the largest gauge will be detected. Moreover, it can be further checked whether the potentially connected tracks are compatible. For example, a meter gauge ego track cannot be connected to a branching normal gauge track and thus, no blade is detected.

[31] A specific frog detector might be used instead, as proposed for example in [Hackel et al. 2015]. However, 3d information is required for this.

[32] Since no distances in y-direction can be observed in the considered 2d lidar sensor measurements, the consistency of the outer rail of the branching track (i.e., the left rail in Figure 2.25b) is ensured when both the inside l_{in} and the outside part l_{out} belong to the same cluster from Algorithm 2.1.

Detection of the branching direction thereon

Idea: When driving again in reading direction in Figure 2.27a, the railway vehicle runs towards position B. In contrast, the railway vehicle would drive towards position C if the setting of the switch blades was as shown in Figure 2.27b. Overall, single turnouts allow four different routes which are called branching directions: facing left (A→B), facing right (A→C), trailing left (B→A), and trailing right (C→A). Facing means that the railway vehicle is moving towards the tip of the frog, i.e., the single turnout is passed from its beginning (A) towards one of both ends (B or C), while trailing qualifies the opposite direction. The sides left and right are always defined in viewing direction from position A. They are independent of the driving direction and whether tracks are curved or not. Furthermore, it is irrespective of whether the branch is located on the left or the ride side (cp. left-hand and right-hand single turnouts in Figure 2.23a and Figure 2.23b respectively). A method for the detection of the branching direction is proposed subsequently, which is motivated by observations from [Rahmig and Johannes 2013].

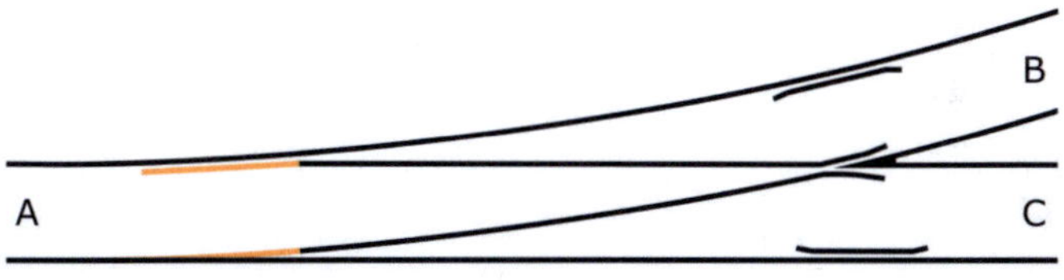

(a) Facing (or trailing) left.

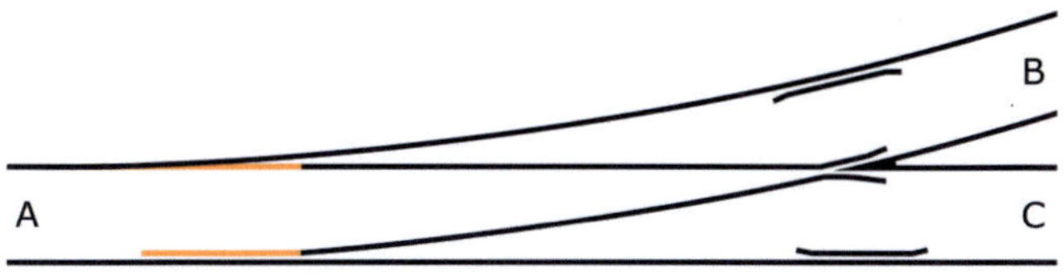

(b) Facing (or trailing) right.

Figure 2.27: Idea of branching directions on single turnouts. Facing characterizes a passing from position A to B or C, while trailing qualifies the opposite direction.

Method: To directly decide on the branching direction after each scan, previously detected frogs and blades can be used (cp. TURNOUTTIPPOSITIONS in Algorithm 2.4). Based on the chronology and the side of their observations, the branching direction can be determined. This is a decision on one out of four possibilities (cp. Table 2.3). The chronology describes whether the branching track is observed inside or outside of the ego track first. Moreover, the branching track can be on the left or the right side of the ego track. Based on these observations, branching events can be raised (cp. Table 2.4 and BRANCHINGEVENTS in Algorithm 2.4). They include the estimated reference position of the blade, the point of its detection, and the detected branching maneuver. Corresponding information on the frog might be included in addition. Finally, even the number of observations can be added. This includes the part of the branching track, which is located completely outside of the ego track and the part, which is partially inside. The latter one is further divided into both halves of the ego track (i.e., the half of the ego track, which includes the tip of the blade, and the opposite side, which is close to the frog). Exemplary branching direction detection results are already shown in Figure 2.26.

Algorithm 2.4: Basic structure of the inference of branching events (detected rail sections C_{all} from Algorithm 2.1, detected track sections T_{all} from Algorithm 2.2, ego, left, and right tracks $E_{\text{all}}, L_{\text{all}}, R_{\text{all}}$ from Algorithm 2.3, sliding window size ω, minimal number of observations ν, number of scans O).

```
procedure BranchingDetection(C_all, T_all, E_all, L_all, R_all, ω, ν)
    F_all ← ∅                                          ▷ all potential frogs
    M_all ← ∅                                          ▷ all potential blades
    B_all ← ∅                                          ▷ all branching events
    for scans s = n^(1) ... n^(O) do
        (F_all, M_all) ← TurnoutTipPositions(s, C_all, T_all, E_all, L_all, R_all, ν)
        B_all ← BranchingEvents(s, F_all, M_all, B_all, C_all, T_all, E_all, L_all, ...
                                        R_all, ω, ν)
    end for
    return B_all
end procedure
```

Table 2.3: Determination of branching directions on single turnouts.

chronology of the branching track relative to the ego track		side of the branching track	branching direction (abbreviation)
first part	**second part**		
inside	outside	left	facing right (FR)
		right	facing left (FL)
outside	inside	left	trailing left (TL)
		right	trailing right (TR)

Remarks: The parameters ω, ν, and $G_{\max}$ are chosen as before. Similar to topology events, points in time are used for the position of the detection and the reference element. The optional number of observations is again an indicator for the certainty of the current event as proposed by [Rahmig et al. 2013]. To achieve (stochastic) independence, every turnout is reported at most once and not before at least ν observations have been made in each of the three parts of the branching track (cp. Table 2.4). Additionally, each potential blade cluster is considered at most once, since it cannot be used within several turnouts on the ego track. To guarantee that the last observation on the blade rail is determined as its tip, it is reported at least ω scans after its latest detection (and thus much later than the tip of the frog).

Extension option for various gauges: Even Algorithm 2.4 can be extended for mixed gauge tracks with three rails, which makes use of the same ideas. As previously discussed, the blade which belongs to the largest gauge of the branching track is detected. It is reported at most once and the compatibility of the connected gauges is checked again. Additionally, the information on the connected gauges that occur on all three sides of the single turnout (cp. positions A, B, and C in Figure 2.27) can be added to the corresponding branching event.

Table 2.4: Contents of a branching event. Optional entries are *italic*.

category	**entry**	**values**	**unit**
points in time for blade	tip position	double	s
	detection	double	s
points in time for frog	*tip position*	*double*	*s*
	detection	*double*	*s*
maneuver	direction	facing | trailing	none
	side	left | right	none
number of observations	*inside, tip of blade*	*unsigned integer*	*none*
	inside, tip of frog	*unsigned integer*	*none*
	outside	*unsigned integer*	*none*

2.6 Interim Conclusion

In summary, a multistage approach has been proposed. It allows to detect the topology of the railway network and the branching direction on single turnouts and bases on the overall structure shown in Figure 2.2. It makes use of the characteristics of a common railway network and follows its structure bottom-up. Motivated by the appearance of railway objects in 2d lidar sensor measurements and their characteristics, features as well as region based and model based approaches are used to detect rail profiles. Based on this, rail and track sections of the ego and nearby tracks can be determined step-by-step. Finally, the position of turnouts and even the branching direction thereon has been detected. In addition, extension options for its use in more complex, but rare scenarios with several rails per track have been proposed. Its implementation will be evaluated experimentally next.

3 Experimental results

In order to validate the proposed approach, various experiments have been performed on different tracks, including several types of turnouts on a commercial railway test ground. A large amount of measurement data has been recorded, which allows to evaluate the results of the proposed multistage approach step by step (cp. also its overall structure in Figure 2.2). For this purpose, the experimental setup, substantial properties of the test ground and the considered data sets are introduced first. Then, the results of the rail profile, rail and track section as well as the topology and branching direction detection are discussed. The evaluation starts with a qualitative analysis of a single data set and a detailed consideration of each step in the corresponding sections. After that, derived quantitative results are presented in a general view. On top of that, Chapter 4 extensively discusses the achieved results, highlights the contribution, and works out the general benefit.

3.1 Overview of the experiments

Based on the proposed approach, the experiments were designed to cover a wide range of topological situations, which are passed at different velocities and also contain a large number of branching situations. The characteristics of the experimental setup, the demanding test ground, and the selected data sets are described in the following.

3.1.1 Experimental setup

As proposed in the setup in Figure 2.3, a lidar sensor[33] has been mounted additionally on its front for the experiments (cp. Figure 3.1). Measuring from a mounting height of $H = 3.97\,\mathrm{m}$ with a field of view of $\alpha = 360°$ and an angular resolution of $\Delta\varphi = {}^{1}/_{14}°$, the setup provides a "panoramic view" with fine resolution in the x-z-plane, while the chosen measurement rate of $f = 50\,{}^{\mathrm{scans}}/_{\mathrm{s}}$ provides dense measurements in y-direction.

Figure 3.1: Experimental railway vehicle IFS 1. The lidar sensor (scale-up on the lower left side) is mounted vertically on the front. Thin orange lines represent selected laser beams which generate a profile of the environment.

For the experiments, the IFS 1 has been used. It is a powered experimental railway vehicle for normal gauge tracks of the RWTH Aachen University. The IFS 1 already has several navigation and motion sensors installed [Lüdicke et al. 2014].

The measurements are restricted to $|x| \leq 6\,\mathrm{m}$ and $|z+H| \leq 0.5\,\mathrm{m}$ as proposed in Section 2.3.1 in order to sense only the relevant railway infrastructure elements.

[33] Pepperl+Fuchs OMD30M-R2000-B23-V1V1D-1L [Pepperl+Fuchs 2016a; Pepperl+Fuchs 2016b]

3.1.2 Test ground

All experiments have been performed in cooperation with the RWTH Aachen University on a commercial test ground [Siemens 2012] in Germany (cp. also Figure 3.2). Its area is set up for both meter and normal gauge railway vehicles and is predominantly built up of Vignole rails. Thus, it is possible to test the normal gauge case which is the most common railway scenario in the world. Additionally, even varied and more complex topologies that are quite rare in real world scenarios are provided thereon (cp. normal gauge and more complex turnouts in Figure 3.3), which allows to test the universality and robustness of the proposed approach. Overall, this area is an ideal and at the same time demanding test ground.

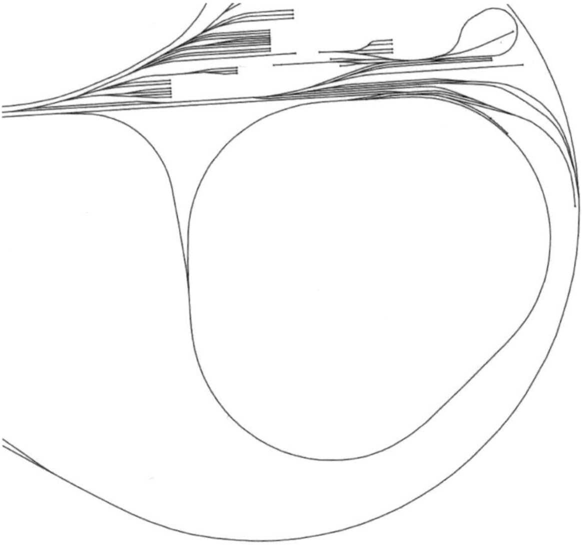

Figure 3.2: Railway network of the commercial test ground (map data: © OpenStreetMap contributors[34]). The map only shows relevant parts where each line represents a track, not a rail.

[34] Contributions from the author see `https://www.openstreetmap.org/user/steindcom/history#map=15/51.1162/6.2213` (accessed: 26.06.2017).

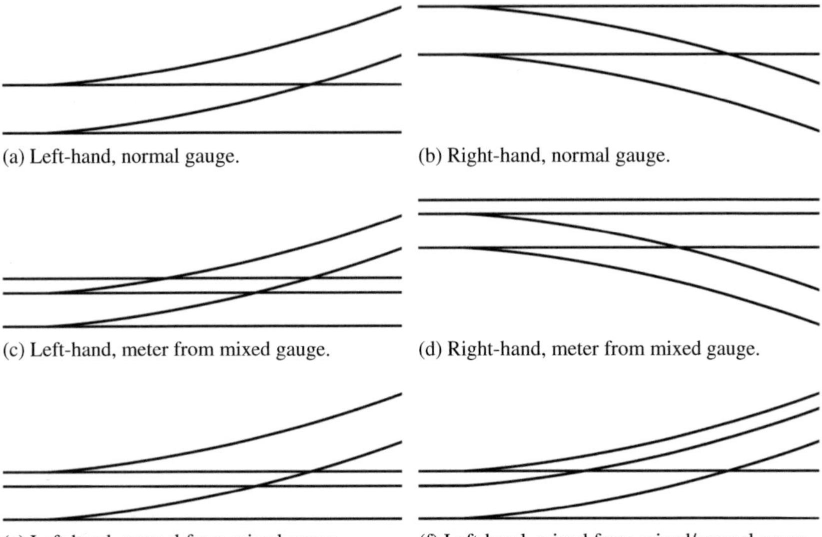

(a) Left-hand, normal gauge. (b) Right-hand, normal gauge. (c) Left-hand, meter from mixed gauge. (d) Right-hand, meter from mixed gauge. (e) Left-hand, normal from mixed gauge. (f) Left-hand, mixed from mixed/normal gauge.

Figure 3.3: Simplified representation of all passed single turnout types.

3.1.3 Data sets

From many kilometers of measurements recorded during day and night, nine data sets with a total length of about 24 km were selected. On the one hand, these data sets contain the repeated passing of the same place at different velocities and in opposed driving directions, which allows to analyze the reproducibility of the results. On the other hand, sections where several turnouts are located closely together reflect typical railway station scenarios in which many branching situations occur. The data sets can be classified in tracks of the circuit as well as siding and exchange tracks. Table 3.1 gives a detailed overview, particularly on the branching situations.

Table 3.1: Characteristics of the nine data sets (maximal velocity v_{max}, sequence of up to nine passed turnouts T*, and branching direction thereon, whereby FL … TR represent facing left … trailing right). Six out of 18 turnouts have been passed in different branching directions. In three out of 56 cases presented in *italic*, the turnouts have not been passed completely.

data set	v_{max}	passed turnouts and branching direction thereon								
		1	2	3	4	5	6	7	8	9
D1	40 km/h	T2 FR	T1 FR	T3 FR	T2 FR	T1 FR	T3 FR	T2 FR	*T1 FR*	
D2	40 km/h	*T1 TR*	T2 TR	T3 TR	T1 TR	T2 TR	T3 TR	T1 TR	T2 TR	
D3	80 km/h	T2 FR	T1 FR	T3 FR	T2 FR	T1 FR	T3 FR	T2 FR	T1 FR	
D4	60 km/h	T2 FR	T1 FR	T3 FR	T2 FR	T1 FR				
D5	20 km/h	T2 FR	T1 FR	T3 FR	T2 FR	T1 FR				
D6	shunting	T1 TR	T2 TR							
D7	shunting	T2 FR	T1 FL	T4 FL	T5 TL					
D8	shunting	*T18 TR*	T5 FR	T6 FR	T7 TR	T8 TR	T9 TL	T10 FR	T11 TR	T12 TL
D9	shunting	T12 FL	T11 FL	T13 FL	T14 FL	T15 FR	T16 FL	T17 FR		

- **Circuit:** Data sets D1–6 contain the repeated driving on a circuit with about 2.5 km length of circumference and a curve radius of 300 m at different velocities [Siemens 2012]. Its mixed gauge track with three rails (cp. also Figure 2.20) contains three single turnouts, whereby turnouts T1 and T3 provide a passage for normal gauge vehicles (cp. Figure 3.3e), while turnout T2 is for meter gauge (cp. Figure 3.3c).

- **Siding and exchange tracks:** Data sets D7–9 include the passing of another 15 single turnouts in shunting mode (max. 25 km/h). They cover common normal gauge single turnouts — both left-hand (T5, T6, T7, T12, T16, and T18 cp. Figure 3.3a) and right-hand (T4, T10, T11, T13, T14, T15, and T17 cp. Figure 3.3b). They also contain further types of single turnouts connecting mixed gauge track with three rails with normal gauge (cp. T8 in Figure 3.3f) or meter gauge tracks (cp. T9 in Figure 3.3d).

Within these data sets, 18 different single turnouts are passed 56 times. In 48 cases, turnouts are passed in facing direction, which leads to ambiguous branching situations. In order to align the number of appearances per branching direction and therefore increase the number of test cases in total, all data sets have been mirrored. By flipping the sign of the x-values in the recorded data and/or reversing them in y-direction, the number of test cases is quadrupled. Thus, an increased number of (virtual) test drives as input sequences allows a varied and extended evaluation of the proposed approach. Table 3.2 illustrates the influence of this mirroring on the branching directions, which are shown in Table 3.1.

Table 3.2: Influence of the mirroring of the data set on the branching direction. FL ... TR represent facing left ... trailing right.

flip in x-direction	no	yes	no	yes
flip in y-direction	no	no	yes	yes
branching directions	FL	FR	TR	TL
	FR	FL	TL	TR
	TL	TR	FR	FL
	TR	TL	FL	FR

3.2 Exemplary evaluation

Each step of the proposed approach contributes to the topology and branching events (cp. also its overall structure in Figure 2.2). Thus, the results of each step are analyzed exemplarily for the shortest data set D6 at first. For their quantitative evaluation, two key performance indicators will be used in particular:

$$R = \frac{\#TP}{\#TP + \#FN} \tag{3.1}$$

$$P = \frac{\#TP}{\#TP + \#FP}. \tag{3.2}$$

The recall R is a measure for the completeness of the detections and the precision P characterizes their accuracy. For both values, $100\,\%$ is the optimal outcome. $\#TP$ denotes the number of correctly detected objects (true positives), while $\#FN$ and $\#FP$ count the number of missing or erroneous detections respectively (false negatives and false positives). $\bar{R}$ and $\bar{P}$ denote arithmetic means of R and P respectively where potential indexes designate the considered data set and/or evaluation step.

The results of the first three *railway infrastructure detection* steps (rail profile, rail section, and track section) described in Chapter 2 are evaluated based on scan-wise labeled areas (ground truth). Therein, one object is a single rail pro-

file in a single scan. Thus, even for this shortest data set, the denominator of R is already a five-digit number.
For each *topology* event, it is evaluated whether the detected neighboring track exists or not. Since only $\#TP$ and $\#FP$ are known, R is abandoned in this evaluation. The denominator of P becomes much smaller, since the events are only raised for neighboring tracks and the events should be independent. For this shortest data set, the denominator of P is 28.
Finally, the detected *turnouts and the branching directions thereon* are evaluated against actual positions of the corresponding tips of both frog and blade as well as the directions the railway vehicle passed them. Thus, it is possible to quantify R and P. Since this shortest data set passes exactly two single turnouts, the denominator of R is two as well.

3.2.1 Detection of rail profiles

Summary

The scan-wise detection of single rail profiles is primarily based on a template matching approach (cp. Section 2.3). As shown in Figure 3.4, the results are almost identical in each case, although the data sets were mirrored. The present rails are almost completely detected, which can be seen from the dominance of black areas in Figure 3.4 representing true positive detections compared to very limited false negatives on turnouts in highly saturated colors. This leads to the recall $\bar{R}_{\mathrm{D6,profile}} = 97.4\,\%$. At this stage, a few false positive detections, shown in brown, arise, which results in the precision $\bar{P}_{\mathrm{D6,profile}} = 85.1\,\%$.

Detailed discussion

Considering the real test drive shown in Figure 3.4c, almost all parts of the rails have been detected successfully. Their positions are — compared to the three mirrored results — quasi congruent (black areas). However, the results cannot be completely identical between all of them, since the RANSAC estimator which is primarily used for the ground detection has a random component.

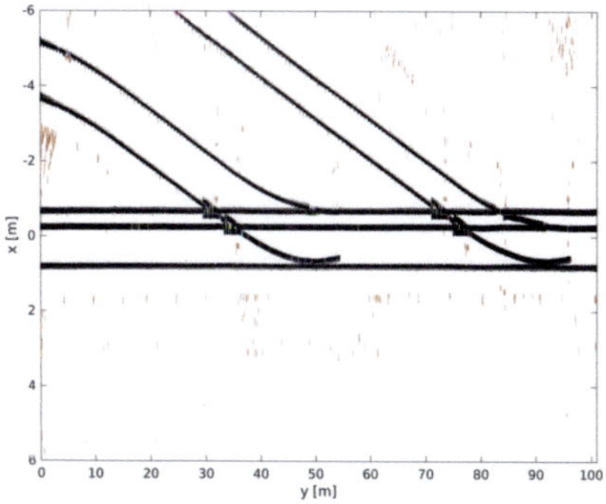

(a) Flipped in x-direction
($R_{\text{profile}} = 97.4\,\%, P_{\text{profile}} = 85.0\,\%$).

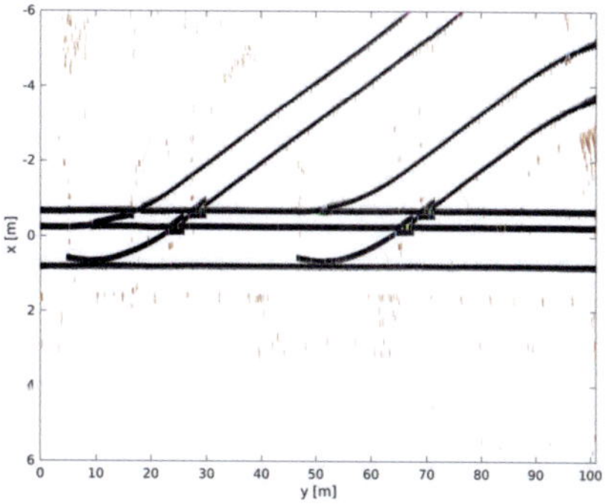

(b) Flipped in x- and y-direction
($R_{\text{profile}} = 97.4\,\%, P_{\text{profile}} = 85.1\,\%$).

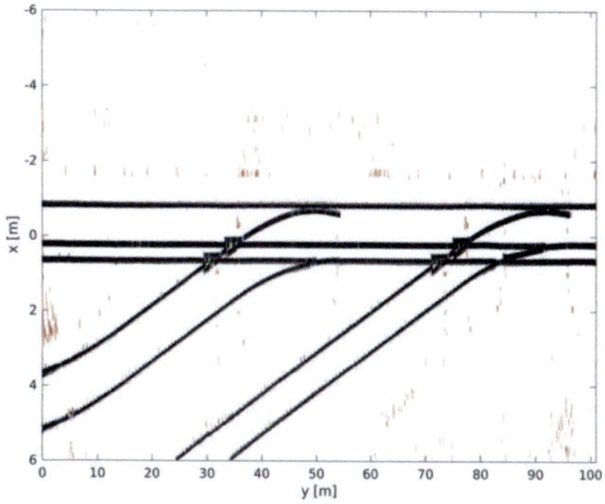

(c) Unflipped
($R_{\text{profile}} = 97.5\,\%, P_{\text{profile}} = 85.2\,\%$).

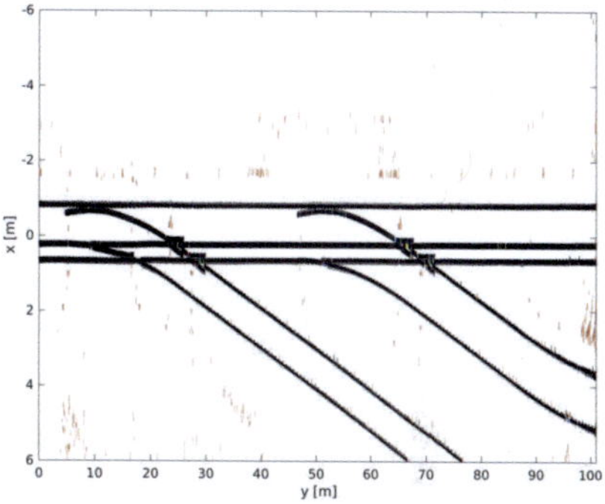

(d) Flipped in y-direction
($R_{\text{profile}} = 97.5\,\%, P_{\text{profile}} = 85.3\,\%$).

Figure 3.4: Rail *profile* detection results for data set D6. In each case, the railway vehicle is (virtually) driving from left to right. True positives are represented in black, false negatives in highly saturated colors, false positives in brown, and white spaces denote both true negatives and regions where no measurements remained after the restriction of the detection area.

Thus, each step can base on a slightly different assignment of measurements to the ground.

The same observations hold for the positions of the false negatives and false positives. The former ones only appear as small gaps near the frogs and on the wing rails (highly saturated colors), while the latter ones are primarily caused by erroneous detections on more or less rail-profile-like objects (brown areas).

Examples of this include power cables ($1.5\,\text{m} \leq x \leq 3\,\text{m}$, $y < 3\,\text{m}$), mechanisms to adjust the positions of the movable frogs (e.g., $1\,\text{m} \leq x \leq 3.5\,\text{m}$, $y \approx 32\,\text{m}$ and $-0.5\,\text{m} \leq x \leq 3\,\text{m}$, $y \approx 36\,\text{m}$ for turnout T1) or the blades ($0\,\text{m} \leq x \leq 2\,\text{m}$, $y \approx 54\,\text{m}$ for turnout T1), or lateral third rails that provide power supply ($x \approx -1.7\,\text{m}$ for the ego track). Most other false positives belong to discontinuities in the environment and appear randomly. However, their spatial extension is limited, so that most of them will be filtered out in the subsequent steps.

Interim conclusion

In other words, more than the existing rail profiles are detected, while the present rails are recognized almost perfectly. This leads to a high completeness of the rail profile detections at the expense of reduced accuracy.

3.2.2 Detection of rail sections

Summary

Based on the rail profile detections presented beforehand, the spatial clustering approach proposed in Section 2.4.1 generates rail sections. The detection results are once again evaluated against the scan-wise labeled ground truth and presented similarly (cp. Figure 3.5). Thus, they are directly comparable to those of the rail profile detection. Although the data sets are mirrored, the related results are quite similar. Generally, almost all true positives remain unchanged. This results in an unaffected recall $\bar{R}_{\text{D6,rail}} = 97.4\,\%$, whereas the number of locally scattered erroneous detections has been slightly reduced, which results in the precision to be increased to $\bar{P}_{\text{D6,rail}} = 87.1\,\%$.

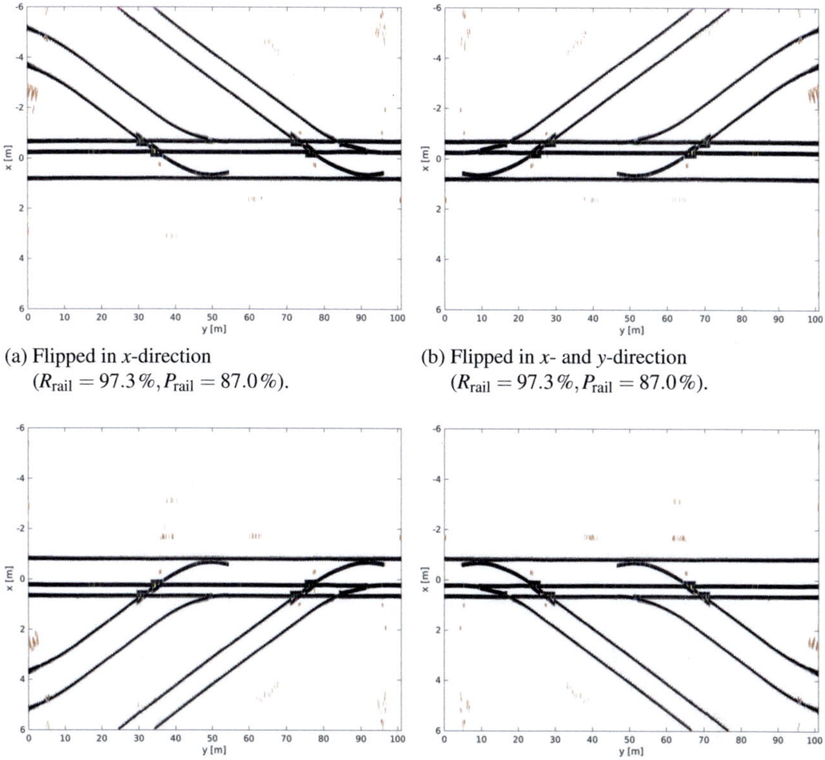

(a) Flipped in x-direction ($R_{\mathrm{rail}} = 97.3\,\%, P_{\mathrm{rail}} = 87.0\,\%$).

(b) Flipped in x- and y-direction ($R_{\mathrm{rail}} = 97.3\,\%, P_{\mathrm{rail}} = 87.0\,\%$).

(c) Unflipped ($R_{\mathrm{rail}} = 97.5\,\%, P_{\mathrm{rail}} = 87.2\,\%$).

(d) Flipped in y-direction ($R_{\mathrm{rail}} = 97.5\,\%, P_{\mathrm{rail}} = 87.3\,\%$).

Figure 3.5: *Rail section* detection results for data set D6. In each case, the railway vehicle is (virtually) driving from left to right. True positives are represented in black, false negatives in highly saturated colors, false positives in brown, and white spaces denote both true negatives and regions where no measurements remained after the restriction of the detection area.

Detailed discussion

All in all, there are no notable changes concerning the true positives and the false negatives, which are drawn in black and highly saturated colors respectively. This demonstrates that the approach is capable of spatial grouping nearby rail detections. However, the position of the remaining false positives shown in brown slightly differs. This is due to the dependence between the direction of clustering and the corresponding normal direction, even though the effects are limited yet. Overall, more or less random detections with few observations can be filtered out, while repeated detections over larger areas cannot yet be eliminated (cp., e.g., the power cables $1.5\,\mathrm{m} \leq x \leq 3\,\mathrm{m}$, $y < 3\,\mathrm{m}$ in Figure 3.5c).

Interim conclusion

The results of the detection of rail sections can be summarized as follows: On the one hand, the spatial clustering slightly increases the accuracy compared to the previous step (cp. the reduced number of brown areas between Figure 3.4 and Figure 3.5). On the other hand, it keeps the almost complete detections, which can again be seen by the dominance of black areas compared to those in highly saturated colors.

3.2.3 Detection of track sections

Summary

The final step in the detection of relevant railway infrastructure elements is the detection of track sections, which has been introduced in Section 2.4.2 and builds up on the previous results. The corresponding results shown in Figure 3.6 are comparable to the previous steps, since they use the same ground truth. The accuracy has been significantly increased to almost perfect $\bar{P}_{\mathrm{D6,track}} = 99.9\,\%$, while the completeness slightly decreased to $\bar{R}_{\mathrm{D6,track}} = 90.7\,\%$.

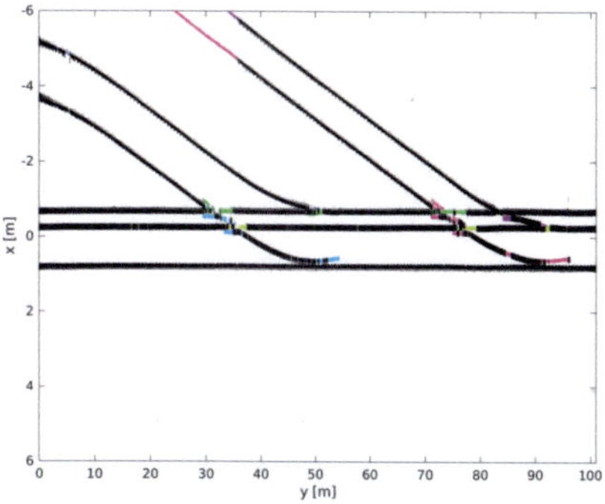

(a) Flipped in x-direction
($R_{\text{track}} = 89.9\,\%, P_{\text{track}} = 99.8\,\%$).

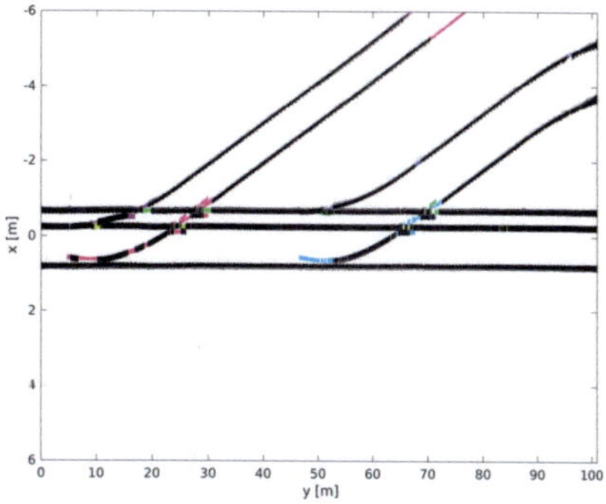

(b) Flipped in x- and y-direction
($R_{\text{track}} = 91.3\,\%, P_{\text{track}} = 100.0\,\%$).

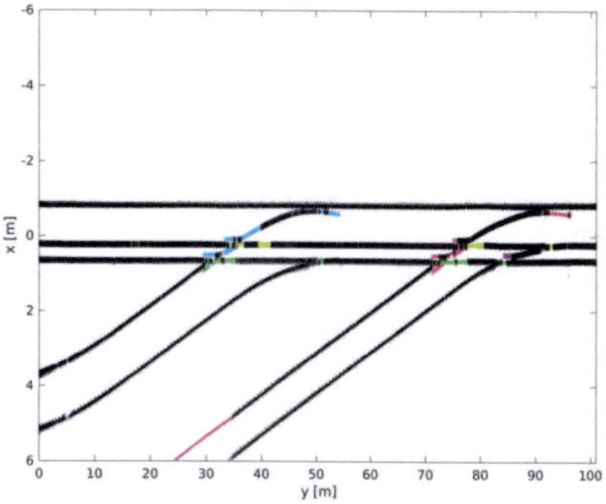

(c) Unflipped
($R_{\text{track}} = 89.6\,\%, P_{\text{track}} = 100.0\,\%$).

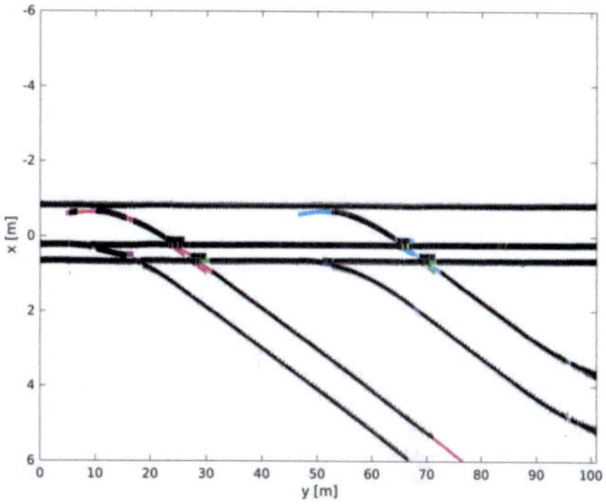

(d) Flipped in y-direction
($R_{\text{track}} = 92.2\,\%, P_{\text{track}} = 100.0\,\%$).

Figure 3.6: *Track* section detection results for data set D6. In each case, the railway vehicle is (virtually) driving from left to right. True positives are represented in black, false negatives in highly saturated colors, false positives in brown, and white spaces denote both true negatives and regions where no measurements remained after the restriction of the detection area.

Detailed discussion

Particularly striking therein is the increase of false negatives. The reasons for the change from black to highly saturated colors are twofold: Firstly, rail sections without parallel rail near the lateral boundaries of the detection area are — depending on the driving direction — not considered, since they can no longer or not yet form a track (cp. for example the light purple rail section near $x \geq 4.8\,\mathrm{m}$, $25\,\mathrm{m} \leq y \leq 35\,\mathrm{m}$ in Figure 3.6c). Thus, the recall for the data sets with flip in y-direction is slightly higher, since the detected branching track can be temporarily continued at the boundary. Secondly, especially on turnouts, large portions of the wing rails (cp. the light blue and dark green areas near $0.1\,\mathrm{m} \leq x \leq 1.0\,\mathrm{m}$, $29\,\mathrm{m} \leq y \leq 36\,\mathrm{m}$ in Figure 3.6c) and some parts of the non-fitting blade rail (cp. the light blue areas near $-0.7\,\mathrm{m} \leq x \leq -0.5\,\mathrm{m}$, $50\,\mathrm{m} \leq y \leq 55\,\mathrm{m}$ in Figure 3.6c) are no longer included in the detections. In all cases, this is correct, since the rails form no track with valid gauge within the detection area on which the railway vehicle can run on. However, the missing corresponding rail sections are treated as false negatives in the performance measures to allow comparisons between the three steps. The remaining false negatives result from temporarily neglecting rail sections when forming track sections. They occur in particular on turnouts, but are more or less random. Remarkable, however, is the fact that hardly any false positives remain, which can be seen by the significant reduction of brown areas.

Interim conclusion

All in all, the track section detection provides highly accurate results with a slight loss on completeness, while the results slightly depend on the direction of the tracks. Their (ir)relevance for the detection of the topology and the branching direction on turnouts is addressed in the subsequent sections.

3.2.4 Detection of topology events

Summary

Based on the detection of track sections presented beforehand, information on their topology can be inferred as proposed in Section 2.5.1. The results shown in Figure 3.7 are — independent of the mirroring — quite similar. Detected events on the left or the right side are shown as vertical purple or green lines respectively, while their start and end points correspond to the detected track centers. False positive detections would be represented by the same color, but with vertical lines not starting or ending in a track center. Since all 28 detected topology events fit the detected tracks even in the visual representation, they are true positives. Finally, this leads to $\bar{P}_{\mathrm{D6,topology}} = 100\,\%$.

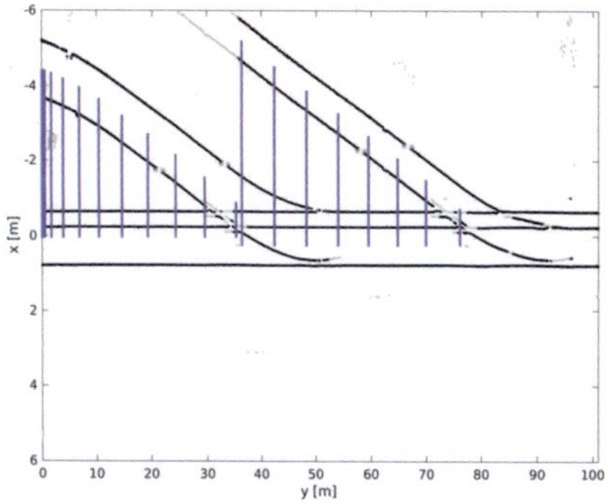

(a) Flipped in x-direction ($P_{\mathrm{topology}} = 100\,\%$).

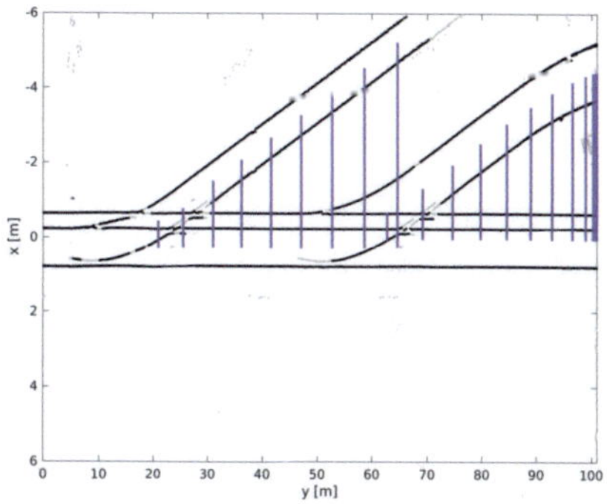

(b) Flipped in x- and y-direction ($P_{\mathrm{topology}} = 100\,\%$).

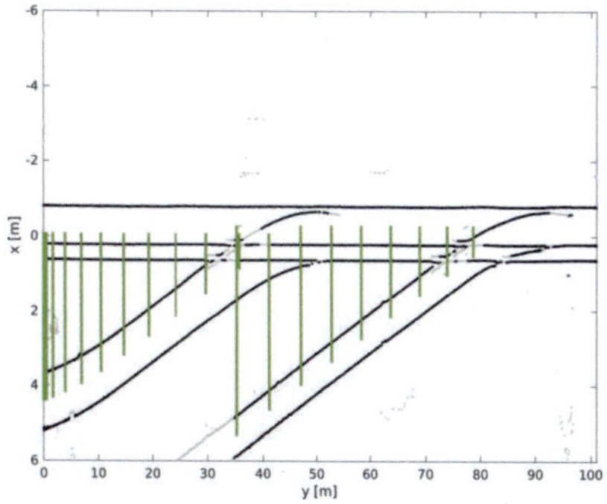

(c) Unflipped ($P_{\mathrm{topology}} = 100\,\%$).

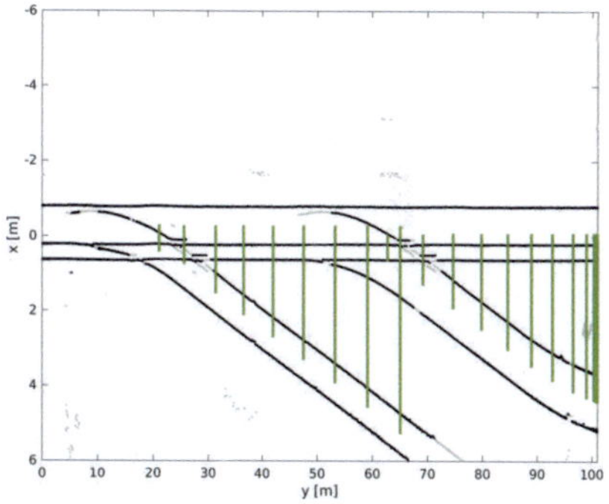

(d) Flipped in y-direction ($P_{\mathrm{topology}} = 100\,\%$).

Figure 3.7: *Topology* detection results for data set D6. In each case, the railway vehicle is (virtually) driving from left to right. Vertical purple lines indicate the position in y-direction where a topology event on the left side has been raised, while their start and end point in x-direction represent the estimated track centers of the other and the ego track. Accordingly, green lines represent detected topology events on the right side. Black dots represent rail centers of detected track sections, whereas gray dots correspond to centers of excluded rail sections.

Detailed discussion

The spacing in y-direction between events on the same side depends on the velocity of the railway vehicle and the constant measurement rate of the lidar sensor. Since the events should be independent from previous ones, there is a minimal space between them. Figure 3.7c shows for example 12 events raised when driving with constant velocity (cp. $15\,\text{m} \leq y \leq 75\,\text{m}$), while several nearby events especially occur in the beginning. Due to the speed-up phase of the railway vehicle, the number of events for the branching normal gauge track ($y \leq 55\,\text{m}$) is higher than for the branching meter gauge track ($y \geq 25\,\text{m}$).

Interim conclusion

All in all, the topology is accurately detected, while only slight temporary differences occur between the mirrored data set.

3.2.5 Detection of the branching direction on turnouts

Summary

Based upon the rail and track section detections, even turnouts and the branching direction thereon can be inferred as proposed in Section 2.5.2. The detection of the passing of both the frog and the blade (and thus the turnout) is shown for the exemplary data set in Figure 3.8 and Figure 3.9 respectively. Each vertical solid line represents a turnout detection, while each vertical dotted line indicates the estimated reference position in y-direction of the corresponding tip of the frog or the blade. The color of the line additionally encodes the detected branching direction. Green is used when passing trailing left, purple when passing trailing right, and red and blue when facing left and right respectively. Since each plot contains two detected turnouts and their line colors correspond to the expected branching directions (cp. Table 3.1 and Table 3.2), all branching directions (and thus the corresponding turnouts) have been detected accurately and completely. This leads to an overall recall and precision of $\bar{R}_{\text{D6,frog}} = \bar{R}_{\text{D6,blade}} = \bar{P}_{\text{D6,frog}} = \bar{P}_{\text{D6,blade}} = 100\,\%$.

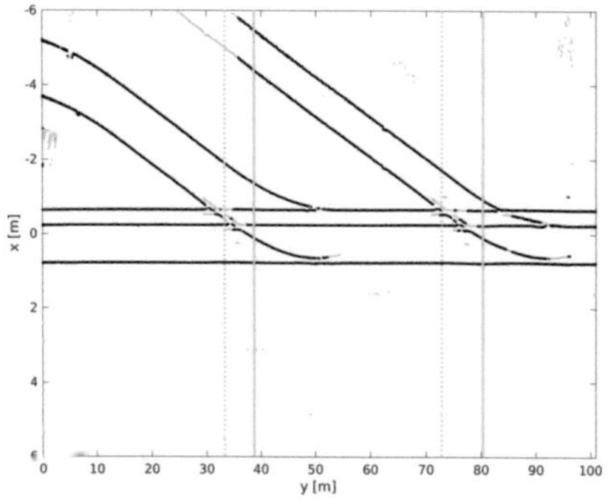

(a) Flipped in x-direction
($R_{\text{frog}} = 100\,\%, P_{\text{frog}} = 100\,\%$).

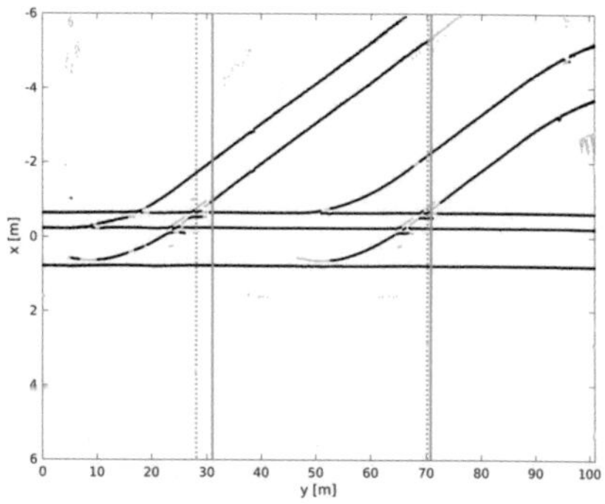

(b) Flipped in x- and y-direction
($R_{\text{frog}} = 100\,\%, P_{\text{frog}} = 100\,\%$).

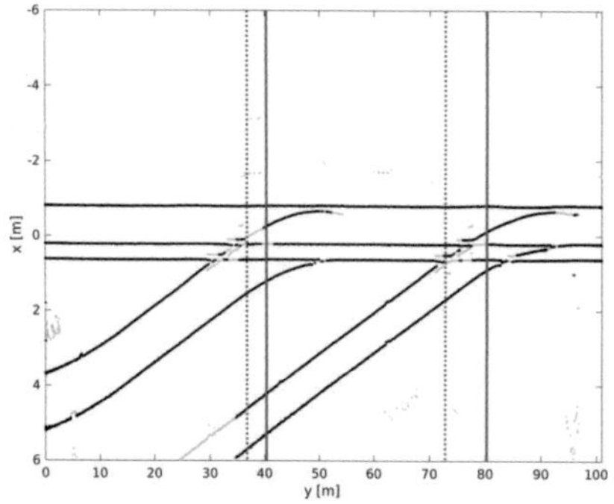

(c) Unflipped
($R_{\text{frog}} = 100\,\%, P_{\text{frog}} = 100\,\%$).

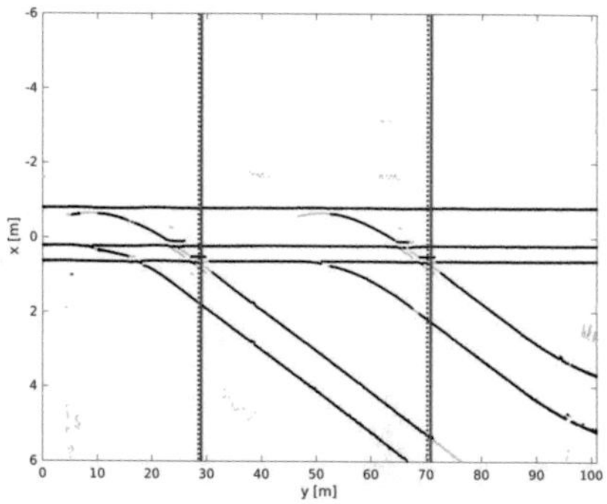

(d) Flipped in y-direction
($R_{\text{frog}} = 100\,\%, P_{\text{frog}} = 100\,\%$).

Figure 3.8: *Frog* based branching direction detection results for data set D6. In each case, the railway vehicle is (virtually) driving from left to right. Vertical solid lines indicate the position where the branching has been detected, while dotted lines represent the position of the estimated tip of the frog. The color of each line indicates the detected branching direction where red represents facing left, blue facing right, green trailing left, and purple trailing right. Black dots represent rail centers of detected track sections, whereas gray dots correspond to centers of excluded rail sections.

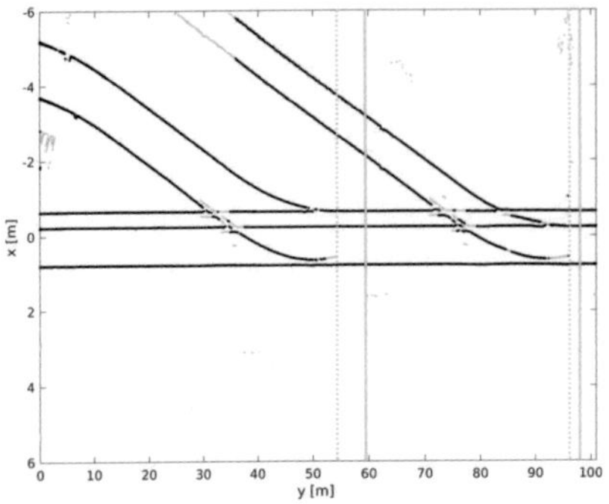

(a) Flipped in x-direction
($R_{\text{blade}} = 100\,\%, P_{\text{blade}} = 100\,\%$).

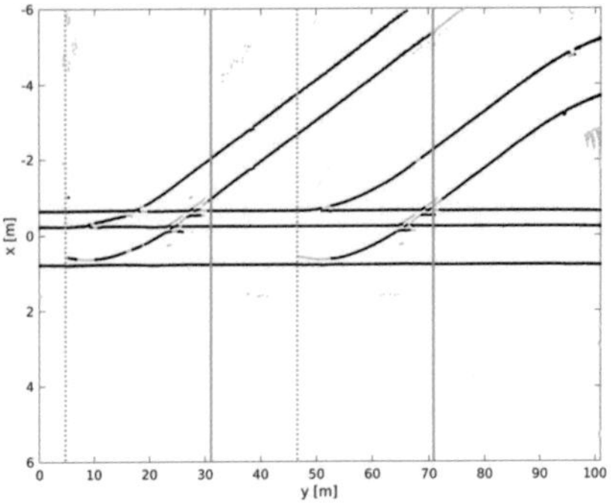

(b) Flipped in x- and y-direction
($R_{\text{blade}} = 100\,\%, P_{\text{blade}} = 100\,\%$).

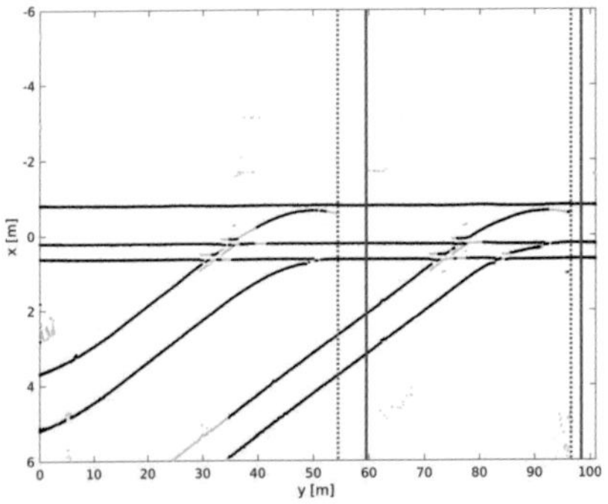

(c) Unflipped
($R_{\text{blade}} = 100\,\%, P_{\text{blade}} = 100\,\%$).

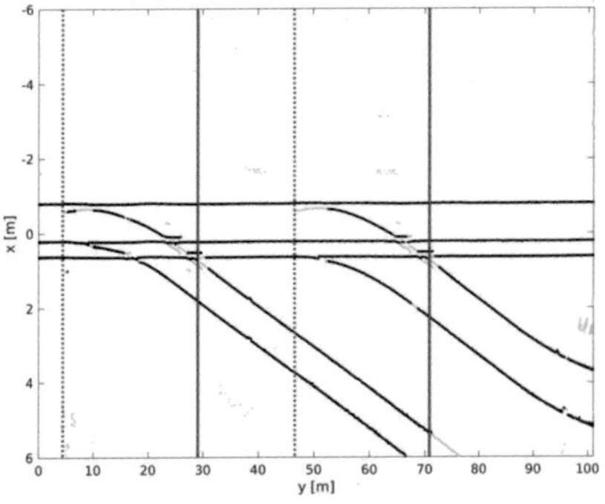

(d) Flipped in y-direction
($R_{\text{blade}} = 100\,\%, P_{\text{blade}} = 100\,\%$).

Figure 3.9: *Blade* based branching direction detection results for data set D6. In each case, the railway vehicle is (virtually) driving from left to right. Vertical solid lines indicate the position where the branching has been detected, while dotted lines represent the position of the estimated tip of the blade. The color of each line indicates the detected branching direction where red represents facing left, blue facing right, green trailing left, and purple trailing right. Black dots represent rail centers of detected track sections, whereas gray dots correspond to centers of excluded rail sections.

Detailed discussion

In each case, the solid lines are located after the passing of the outermost frog, while the positions of the tip of the frog and the blade respectively are estimated opposite to the driving direction. The position of the tips is more important

thereby. The tips of the blades are detected almost perfectly (cp. Figure 3.9), while the estimated position of the frogs differs by a few meters, especially when passing the turnouts in trailing direction (cp. Figure 3.8a and Figure 3.8c). The reason for the latter deviation is that at each single turnout in data set D6, two frogs are passed — one on the meter gauge part and another one on the normal gauge part of the ego track (cp. for turnout T1 its tips at $x = 0.7\,\mathrm{m}$, $y = 31\,\mathrm{m}$ when intersecting the right normal gauge rail and at $x = 0.3\,\mathrm{m}$, $y = 36\,\mathrm{m}$ the right meter gauge rail respectively in Figure 3.8c). However, single turnouts in common railway scenarios only have one frog and do thus not provide such ambiguities.
In general, the space in y-direction between the estimated tip and the detection itself differs between trailing and facing passings. In the case of the frog, this space tends to be larger for trailing passings, since the branching track needs to pass the frog and about half of the gauge of the ego track. The behavior for the blade is vice versa, which also results from the typical dimensions of turnouts. The space between the estimated tip of the blade and the detection itself is larger for facing passings, since the passing of the blades and the frog is required in every case. In the particular case of the blade detection in trailing direction, a constant temporal delay occurs that complies with the sliding window size of 1 s.

Interim conclusion

All in all, it is possible to detect both turnouts and the branching direction thereon, whereby the timings differ between facing and trailing passings. In the considered data set D6, all turnouts and branching directions have been detected accurately and completely.

3.3 Quantitative evaluation

After the exemplary analysis of one data set in detail, the following sections present the results over all data sets and the corresponding conclusions more compactly. The five main steps are considered once more, but a specific mirroring of a data set is only examined when necessary. Derived values of the key performance indicators are considered herein.

3.3.1 Rail profile, rail section, and track section detection

Summary

Since the rail profile, the rail section, and the track section detection are evaluated on the same ground truth, these three steps are analyzed as one. Their results are plotted in Figure 3.10 where each color represents a data set, while the corresponding sequences develop from the left to the right.

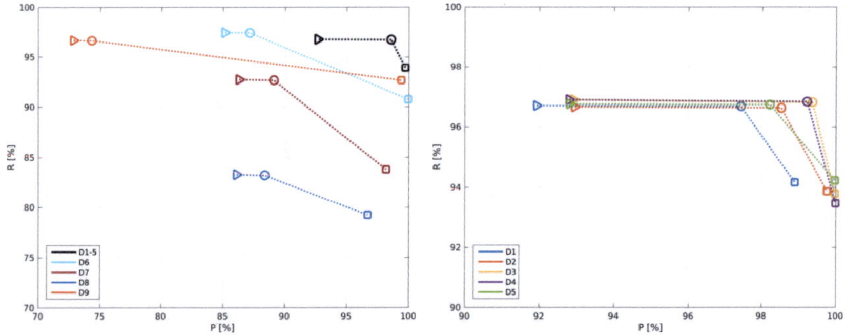

(a) All data sets. (b) Only repeated driving on circuit.

Figure 3.10: Overall results of the rail profile, rail section, and track section detection. Mean values of recall R and precision P per data set, where $\triangleright$ represents results of the rail profile detection, $\bigcirc$ results of the rail section detection, and $\square$ of the track section detection, while the dotted lines are plotted for faster visual assignment. The mean results of data sets D1–5 are summarized in black in (a), whereas (b) shows their individual results in an enlarged view. Both axes in (a) start at 70 % and in (b) at 90 %, while all end at 100 %.

The similar characteristics for each data set are particularly striking. Starting with the most complete results after the rail profile detection in Section 2.3, the detection of rail sections from Section 2.4.1 slightly improves the precision, but hardly changes the recall. Finally, the track section detection proposed in

Section 2.4.2 significantly increases the precision at the expense of a reduced recall.

The detection results improve step by step for each data set. The growth in accuracy is (much) higher than the loss in completeness ($\bar{P}_{\text{track}} - \bar{P}_{\text{profile}} \gg |\bar{R}_{\text{track}} - \bar{R}_{\text{profile}}|$). Furthermore, the detection of track sections concludes with the highest accuracy in every case ($\bar{P}_{\text{track}} \gg 95\,\%$).

Detailed discussion

Within this evaluation, each particular case is assessed regarding the correctness of a single rail profile detection in a single scan, which makes the results of the three different steps comparable. Two observations summarize the results shown in Figure 3.10:

- The values of both key performance indicators differ only slightly within every step between the four mirrored versions of each data set (mean deviation values: 0.3 % for R and 0.1 % for P). Thus, they are represented by a mean value per data set each time. Using the example of data set D6, the four recall and the four precision values denoted in Figure 3.4 are averaged in each case and depicted as light blue ▷ in Figure 3.10a. The same holds for the values of the rail section detection shown in Figure 3.5 and the track section detection from Figure 3.6 that are denoted as light blue ○ and □ in Figure 3.10a respectively.
- A repeated passing of the same location leads to similar detection results. This can be observed in particular from data sets D1–5, which have been recorded within the circuit. Their almost identical results are further summarized in Figure 3.10a, while Figure 3.10b shows an enlarged view. In contrast, varying environments lead to different disturbances, so that the results of the other data sets are not directly comparable.

Occluded rail profiles on level crossings and extended objects with rail like shape have been identified as the major cause for false negative and false positive detections respectively. The data sets with the lowest initial recall and precision are used for their further examination hereinafter.

In data set D8, several level crossings are passed that allow road vehicles to cross the track on the same level. Since the rail shape is not fully visible for the lidar sensor in these regions, they are temporarily not detected as rail profile. Their missing detections are treated as false negatives[35] (cp. the longest level crossings on the ego and the parallel track which are shown in highly saturated colors at $930\,\text{m} \leq y \leq 1015\,\text{m}$ in Figure 3.11) which results in lower recall values.

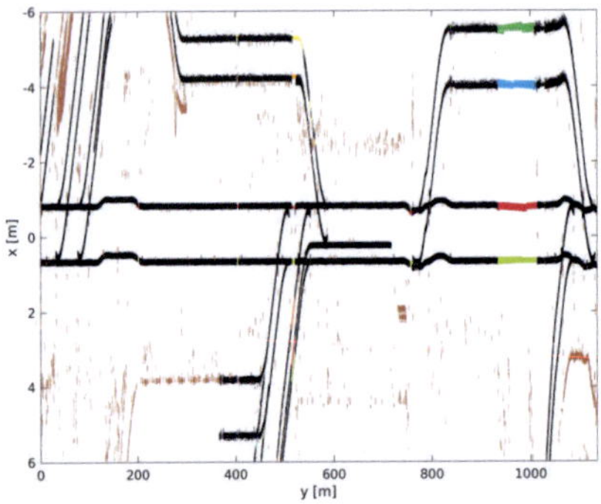

(a) Rail profile detection ($R_{\text{profile}} = 83.3\,\%, P_{\text{profile}} = 86.1\,\%$).

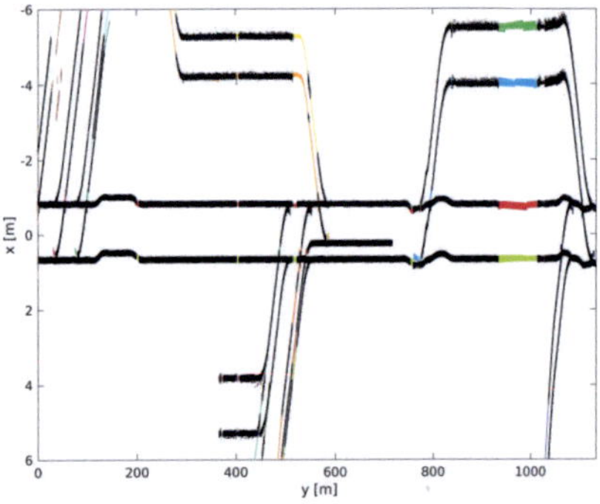

(b) Track section detection ($R_{\text{track}} = 79.2\,\%, P_{\text{track}} = 96.9\,\%$).

Figure 3.11: Detection results for data set D8 (unflipped). The railway vehicle is driving from left to right. True positives are represented in black, false negatives in highly saturated colors, false positives in brown, and white spaces denote both true negatives and regions where no measurements remained after the restriction of the detection area.

[35] Herein, rails are assumed to be detectable when the corresponding track section can be driven on without collision. On the one hand, missing detections that are caused by erroneous assumptions in the proposed approach are treated as false negatives (e.g., the temporary violation of the unique rail profile assumption that cannot be fulfilled on level crossings). On the other hand, rails and tracks that are occluded by other railway vehicles are not detectable by the proposed setup, since the laser beam can neither penetrate objects, nor look behind them. Thus, these sections are not treated as missing detections, but can be temporarily treated as an erroneous detection (cp. the temporarily detected left rail of the partially occluded track at $x \geq 3.8\,\text{m}$, $175\,\text{m} \leq y \leq 365\,\text{m}$ in Figure 3.11a).

The cause for the initially lower precision of data set D9 lies in more or less randomly distributed false positives. They result from erroneous rail profile detections (cp. the brown course caused by vegetation and a wire mesh fence at $x < -2.5\,\text{m}$, $0\,\text{m} \leq y \leq 315\,\text{m}$ in Figure 3.12a) and lead to a reduced precision. However, most erroneous detections can be eliminated in the track section detection. This is evident from Figure 3.12b where the corresponding brown areas have almost entirely disappeared and from Figure 3.10a where the orange □ represents a significantly increased precision.

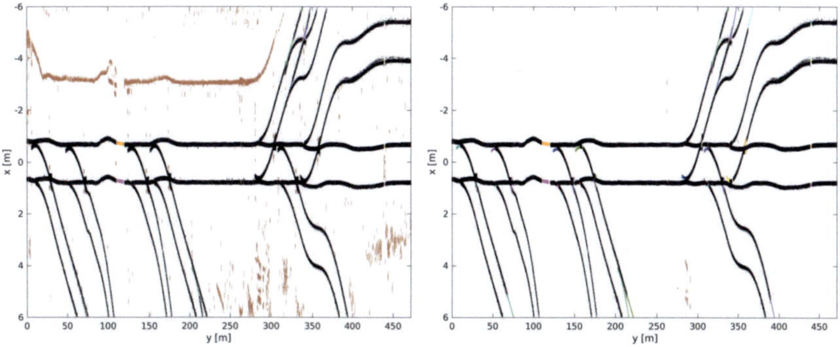

(a) Rail profile detection ($R_{\text{profile}} = 96.7\,\%, P_{\text{profile}} = 72.8\,\%$).

(b) Track section detection ($R_{\text{track}} = 93.1\,\%, P_{\text{track}} = 99.5\,\%$).

Figure 3.12: Detection results for data set D9 (unflipped). The railway vehicle is driving from left to right. True positives are represented in black, false negatives in highly saturated colors, false positives in brown, and white spaces denote both true negatives and regions where no measurements remained after the restriction of the detection area.

Whether and to what extent these results affect the topology detection and especially the detection of turnouts and the branching direction thereon, is analyzed in the subsequent sections.

Interim conclusion

All in all, the results show a similar development over the three considered evaluation steps, while the quantitative values differ due to varying environments. Both the qualitative and the quantitative results demonstrate that the track section detection, reliably and independent of the data set, eliminates almost every false positive detection on the one hand, which results in highest accuracy values up to 100 %. On the other hand, the vast majority of rails is successfully detected, whereby the loss in completeness between the rail profile detection and the track section detection is less important than the increase of accuracy. In the end, less than every 120th detection is erroneous, while the detections have a completeness of 79 to 95 %.

3.3.2 Detection of topology events

Summary

The topology of the railway network within the nearby environment can be inferred from the previously detected track sections. For each corresponding event, it has been manually evaluated whether it fits to an actually existing track or not. Thus, only the correctness of the results can be specified by the precision $\bar{P}_{\text{topology}} = 98.5\,\%$ (cp. Table 3.3). Thereby, no significant deviations occur between individual data sets or their mirrorings. Overall, the vast majority of neighboring tracks has been detected correctly — regardless of their side.

Table 3.3: Overall results of the topology detection. $\bar{P}_{\text{topology}}$ denotes the mean precision for the detected topology events (i.e., track on the left or right side) and $\#TP$ and $\#FP$ count the number of correct and erroneous detections respectively.

side	$\#TP$	$\#FP$	$\bar{P}_{\text{topology}}$
left	1135	17	98.5 %
right	1149	18	98.5 %
any	2284	35	98.5 %

Detailed discussion

It can be observed that three main causes affect the topology detection results. The velocity of the railway vehicle has the largest influence on the spacing in

y-direction between topology events on the same side. Moreover, both types of erroneous track section detections discussed in Section 3.3.1 have an impact. On the one hand, false negative track section detections increase the distance between consecutive topology events, while they do not affect the correctness. On the other hand, false positive track section detections can result in erroneous topology events. Those effects are illustrated using data set D7, which provides 68 correct topology detections besides the maximal number of erroneous ones.

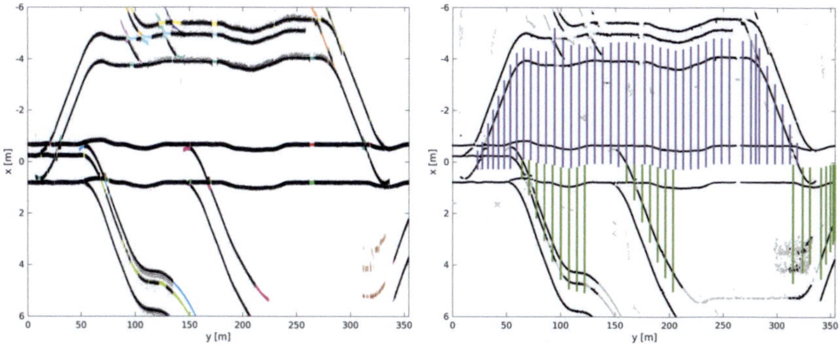

(a) Track section detection ($R_{\text{track}} = 84.3\,\%, P_{\text{track}} = 98.3\,\%$).

(b) Topology detection ($P_{\text{topology}} = 95.8\,\%$).

Figure 3.13: Detection results for data set D7 (unflipped). The railway vehicle is driving from left to right.
(a): True positives are represented in black, false negatives in highly saturated colors, false positives in brown, and white spaces denote both true negatives and regions where no measurements remained after the restriction of the detection area.
(b): Vertical purple lines indicate the position in y-direction where a topology event on the left side has been raised, while their start and end point in x-direction represent the estimated track centers of the ego track and the other track. Accordingly, green lines represent detected topology events on the right side. Black dots represent rail centers of detected track sections, whereas gray dots correspond to centers of excluded rail sections.

In Figure 3.13b, a vertical purple or green line is drawn each time when a topology event has been raised on the left or the right side respectively. It shows that almost every detected neighboring track visually matches the actually existing ones. Moreover, the start and end points of each line correspond to the track centers detected for the ego and the neighboring track. The latest topology

events are much closer to each other, since the railway vehicle slowed down (cp. the green lines for detections on the right side at $y > 340$ m in Figure 3.13b).
Partially missing track section detections result from two level crossings (cp. the areas in highly saturated colors on the left parallel track at $x < -3.8$ m, $170\,\text{m} < y < 180\,\text{m}$ and $260\,\text{m} < y < 270\,\text{m}$ in Figure 3.13a). They result in a slightly larger spatial distance between the corresponding topology events (cp. the increased spacing between successive purple lines near $y = 175$ m and $y = 265$ m in Figure 3.13b). Since missing detections cannot result in any topology detection, they do not influence the correctness of the detected topology.
In contrast, extended areas with false positive detections of track sections may cause erroneous topology detections. For $205\,\text{m} \leq y \leq 340\,\text{m}$, no neighboring track can be detected on the right side, which is perfectly correct. At first, its right rail is located outside of the considered detection area. Consecutively, another railway vehicle occupies the track. However, within a certain area, the left rail of this track has been detected again, while some of these detections are associated with erroneous ones to form a track section (cp. the false positive track section detections at $x > 2.8\,\text{m}, 310\,\text{m} < y < 335\,\text{m}$ shown in brown in Figure 3.13a and the corresponding topology events indicated by three green lines at $315\,\text{m} < y < 335\,\text{m}$ in Figure 3.13b). These detections are objectively wrong, even though a neighboring track on the right side exists.
After all, it is noteworthy that even the three tracks branching into the left parallel track have been detected correctly (cp. the purple lines near $y = 95$ m, $y = 132$ m, and $y = 281$ m in Figure 3.13b which point further upwards in each case).

Interim conclusion

All in all, the topology detection shows a good coverage of the neighboring tracks. Nonetheless, these cannot be detected if they are too far away from the ego track or if they are partially hidden, e.g., by another railway vehicle.

However, more than 95 % of the detections are correct. Thus, they provide valuable information on the nearby environment.

3.3.3 Detection of the branching direction on turnouts

Summary

On the basis of the mirroring, every single turnout on the ego track has been passed (virtually) in every possible branching direction (cp. Table 3.1 and Table 3.2). Each of the 212 expected entire passings of single turnouts has been rated, in terms of whether the branching direction (and thus implicitly the turnout itself) has been detected and whether the detected branching direction fits to the expected one. In summary, the vast majority of branching directions has been detected, while almost every detection is accurate (cp. Table 3.4). This results in identical mean values of 98.1 % for the recall and the precision. Moreover, it makes no difference for the values if the detection of the single turnout or the branching direction thereon is considered or whether it is based on the detection of the frog or the blade.

Table 3.4: Overall results of the branching direction detection on turnouts. $\bar{R}_{\text{branching}}$ and $\bar{P}_{\text{branching}}$ denotes the mean recall and precision for the detected branching direction (facing left ... trailing right) and $\#TP$, $\#FN$, and $\#FP$ count the number of correct, missing, and erroneous detections respectively.

branching direction	$\#TP$	$\#FN$	$\#FP$	$\bar{R}_{\text{branching}}$	$\bar{P}_{\text{branching}}$
facing left	52	1	0	98.1 %	100 %
facing right	52	1	1	98.1 %	98.1 %
trailing left	52	1	2	98.1 %	96.3 %
trailing right	52	1	1	98.1 %	98.1 %
any	208	4	4	98.1 %	98.1 %

Detailed discussion

In order to integrate the branching events into a train-borne localization approach later on, two aspects need to be addressed. At first, influences on the correctness and the completeness of branching direction detections are presented. Afterwards, their position accuracy is evaluated.

Correctness and completeness: As already noted, there is no need to distinguish between the correctness of the turnout detections, the correctness of the branching direction detections, or whether they are based on the frog or the blade as reference element, since all of them are equal. This also holds for the completeness. In every case, there is a common major reason for erroneous or missing branching direction detections. On the one hand, false positive detections primarily result from the higher complexity of turnouts, which operate on both meter and normal gauge. On the other hand, false negative branching direction detections often arise from missing track section detections, which are caused by temporarily occluded rails or missing data (cp. Section 3.3.1). Both effects are further analyzed using appropriate examples. In the corresponding diagrams, the y-position of the turnout detection and the detected tip of the frog or the blade are represented by a solid or dotted vertical line respectively. Additionally, the color of each line complies with the detected branching direction. Figure 3.14 shows two detections (cp. both nearby blue solid lines at $y = 73\,\mathrm{m}$) on the second single turnout, which is of a complex type (cp. Figure 3.3e). The tip of the normal gauge blade has been detected correctly (cp. the blue dotted line near $y = 50\,\mathrm{m}$ in Figure 3.14b), while the begin of the meter gauge rail has been detected as the tip of another blade (cp. the blue dotted line at $y = 0\,\mathrm{m}$ in Figure 3.14b). The latter one is incorrect, since it is not a movable rail. However, such types of turnouts are unusual in common railway networks. All other branchings are detected correctly and even match the actual positions of the frog and the blade.

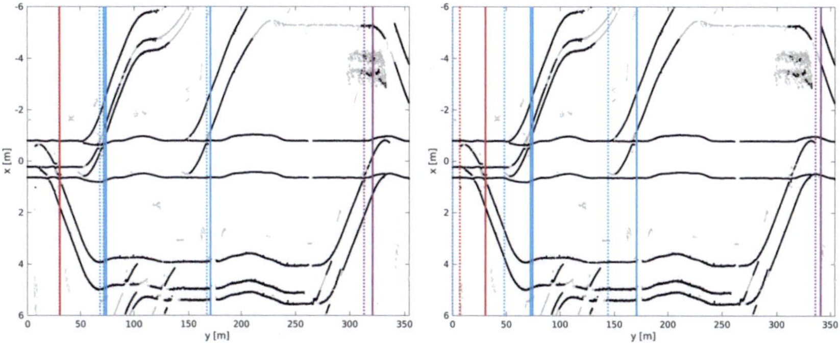

(a) Frog detection ($R_{\text{frog}} = 100\,\%, P_{\text{frog}} = 80\,\%$).

(b) Blade detection ($R_{\text{blade}} = 100\,\%, P_{\text{blade}} = 80\,\%$).

Figure 3.14: Branching direction detection results for data set D7 (flipped in x-direction). The railway vehicle is (virtually) driving from left to right. Vertical solid lines indicate the position where the branching has been detected, while dotted lines represent the position of the estimated tip of the frog or blade respectively. The color of each line indicates the detected branching direction where red represents facing left, blue facing right, green trailing left, and purple trailing right. Black dots represent rail centers of detected track sections, whereas gray dots correspond to centers of excluded rail sections.

Another example in Figure 3.15 shows that all seven branching directions are detected correctly, but a blue solid line near $y = 570\,\text{m}$ representing a facing right branching is missing (which can be better recognized in Figure 3.15a). Buried rails (most probably due to construction work) on the branching track were partially not detectable, so that sections of this track and thus a branching direction has not been detected. In this case, a branching to the left side might lead to a derailment, where its missing detection becomes negligible.

On top of that, it needs to be noted that false positive track section detections did not influence the detection of turnouts. Moreover, all branching events that have been evaluated as false positives were basically correct. They detected the correct branching direction, but the branching itself has been reported for the second time or the turnout has not been passed completely in the data set.

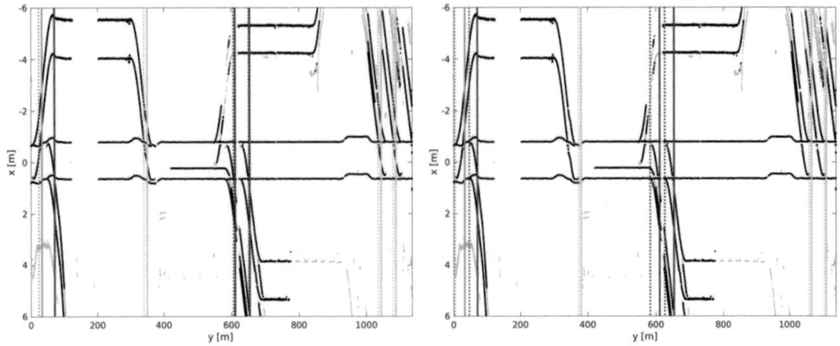

(a) Frog detection
($R_{\text{frog}} = 87.5, \%, P_{\text{frog}} = 100\,\%$).

(b) Blade detection
($R_{\text{blade}} = 87.5\,\%, P_{\text{blade}} = 100\,\%$).

Figure 3.15: Branching direction detection results for data set D8 (flipped in y-direction). The railway vehicle is (virtually) driving from left to right. Vertical solid lines indicate the position where the branching has been detected, while dotted lines represent the position of the estimated tip of the frog or blade respectively. The color of each line indicates the detected branching direction where red represents facing left, blue facing right, green trailing left, and purple trailing right. Black dots represent rail centers of detected track sections, whereas gray dots correspond to centers of excluded rail sections.

Overall, the total number of missing or erroneous detections is low, while no error appeared on normal gauge single turnouts. Thus, it is expected that the number of false detections is even lower on common railway networks.

Position accuracy: The evaluation of the position where the branching direction has been detected, and the estimated reference of all 208 correctly detected single turnouts, shows two things. Firstly, no major differences appear between the left and the right side of a facing or trailing passing. Therefore, only the facing and the trailing direction are considered subsequently. Secondly, though, the results vary between the two detected reference elements. For this, two distance measures and a temporal delay have been determined for both the frog and the blade as shown in Figure 3.16.

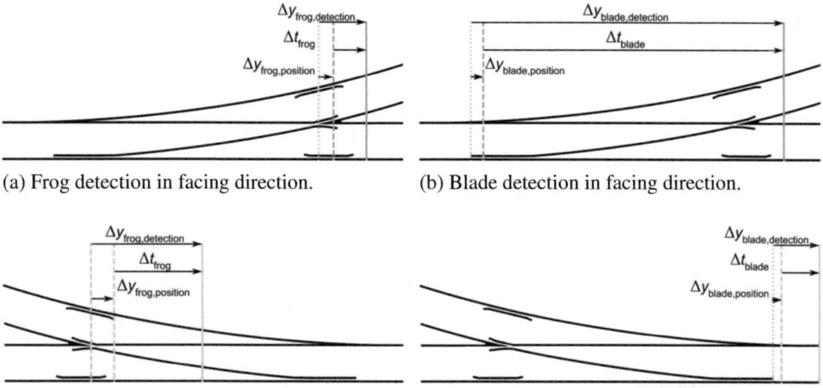

(a) Frog detection in facing direction.

(b) Blade detection in facing direction.

(c) Frog detection in trailing direction.

(d) Blade detection in trailing direction.

Figure 3.16: Evaluation of the detections on single turnouts. Δy denotes a spatial distance, while Δt complies with a temporal delay. Vertical solid lines indicate the position where the branching has been detected. Dashed lines represent the position of the estimated tip of frog or blade respectively, whereas their actual position in y-direction is shown as dotted orange line. Blue and green represent a passing facing right or trailing left respectively.

The position of the tip of the frog and the blade is almost perfectly estimated. This can be seen from Figure 3.17 where blue bars represent the histogram of the deviations from their actual position. The corresponding probability distribution shown in red has its center close to 0 m in every case, but different spreadings. The blade is determined even more precisely where less than 10 % of the deviations are greater than ± 0.5 m. This results in a mean of almost 0 m and a standard deviation of less than 1.5 m (cp. Figure 3.17b and Figure 3.17d). The few significant outliers have two main causes. On the one hand, erroneous detections of the tip of the frog on more complex single turnouts as shown in Figure 3.8c result in an increased deviation (cp. the bar at 6 m representing 4 % of all cases in Figure 3.17c). On the other hand, several missing scans on the blade prevented the approach from estimating the exact position of its tip (cp. the bar at 10 m representing 2 % of all cases in Figure 3.17b). However, their deviation can be described by a normal distribution for the facing and the trailing direction as shown in Figure 3.17. Moreover, in most cases, both the tip of the frog and the blade have been detected with only a small deviation from the exact position.

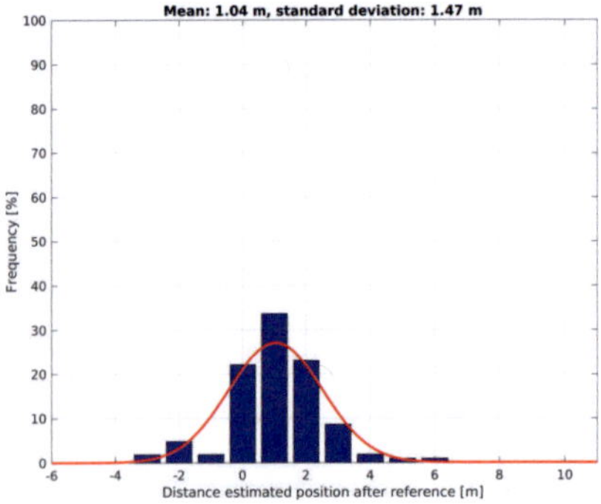

(a) Frog position in facing direction.

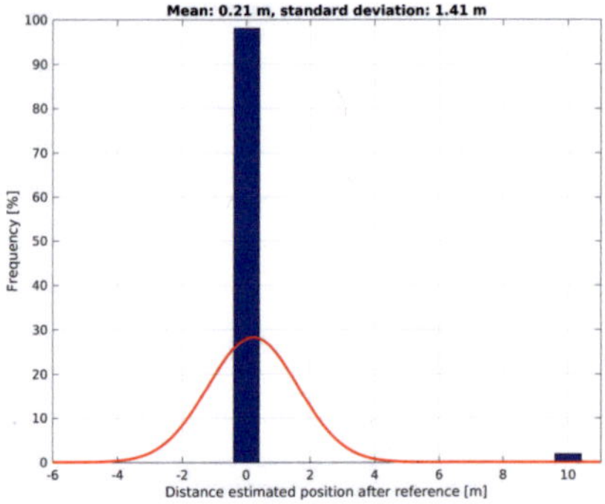

(b) Blade position in facing direction.

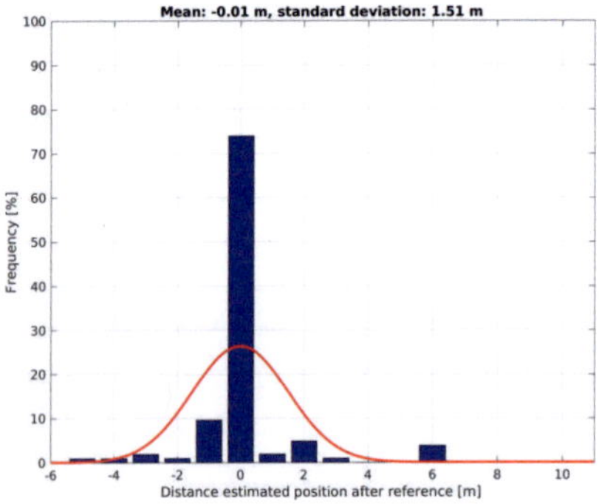

(c) Frog position in trailing direction.

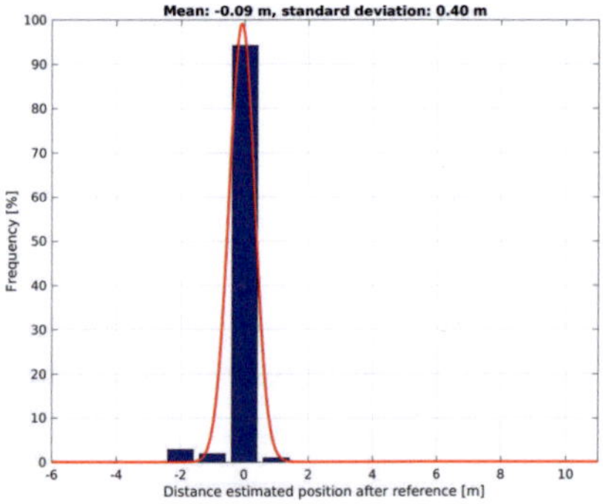

(d) Blade position in trailing direction.

Figure 3.17: Deviation of the estimated reference position from their actual position on single turnouts. The histograms have a bin size of 1 meter. The distribution of $\Delta y_{\text{frog, position}}$ and $\Delta y_{\text{blade, position}}$ respectively is shown in blue, while the red curve represents the corresponding probability distribution, whose mean and standard deviation are denoted above each figure.

The differences in the spatial distance between the tip of the reference element and its detection (cp. the four plots in Figure 3.18) and the temporal delay between the estimated tip of the reference element and its detection (cp. the four plots in Figure 3.19) primarily result from the dimensions of the turnout and the velocity of the railway vehicle. Once more, histograms with integer resolution and the corresponding probability distribution are shown.

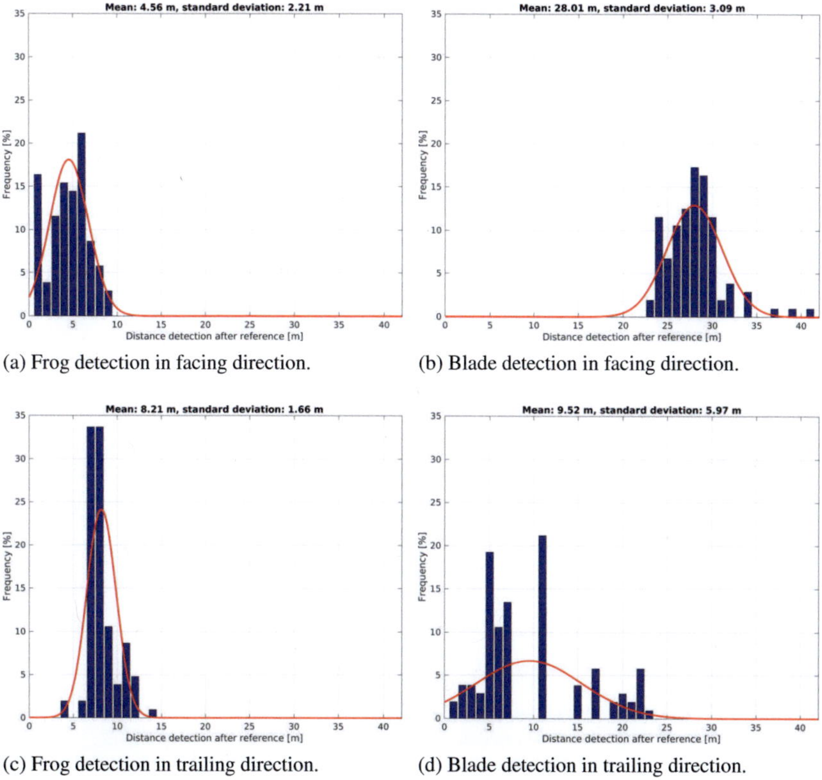

(a) Frog detection in facing direction.

(b) Blade detection in facing direction.

(c) Frog detection in trailing direction.

(d) Blade detection in trailing direction.

Figure 3.18: Distance between the actual reference position and their detection on single turnouts. The histograms have a bin size of 1 meter. The distribution of $\Delta y_{\text{frog, detection}}$ and $\Delta y_{\text{blade, detection}}$ respectively is shown in blue, while the red curve represents the corresponding probability distribution, whose mean and standard deviation are denoted above each figure.

The distribution of the spatial distance between the tip of the blade and its detection in facing direction (cp. Figure 3.18b) highly resembles those of the frog (cp. Figure 3.18a). The major difference is the constant shift to the right side, which results from the distance between the tip of the blade and the tip of the frog.

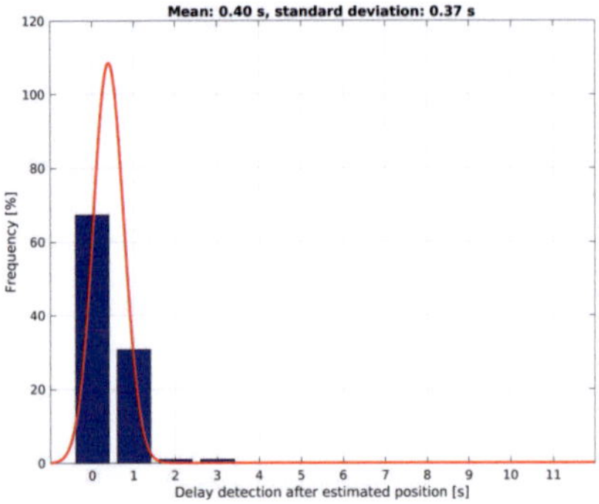

(a) Frog delay in facing direction.

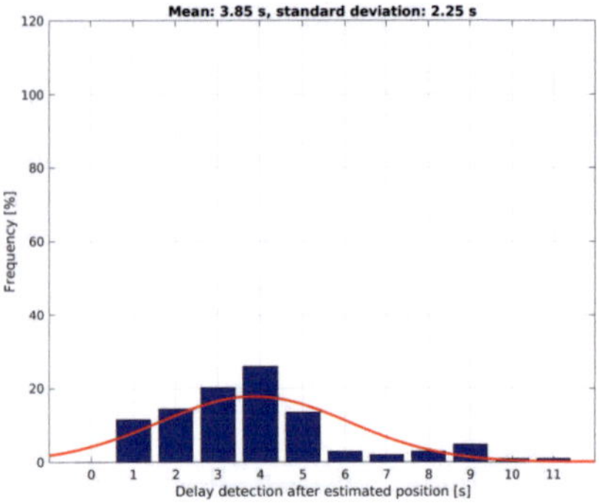

(b) Blade delay in facing direction.

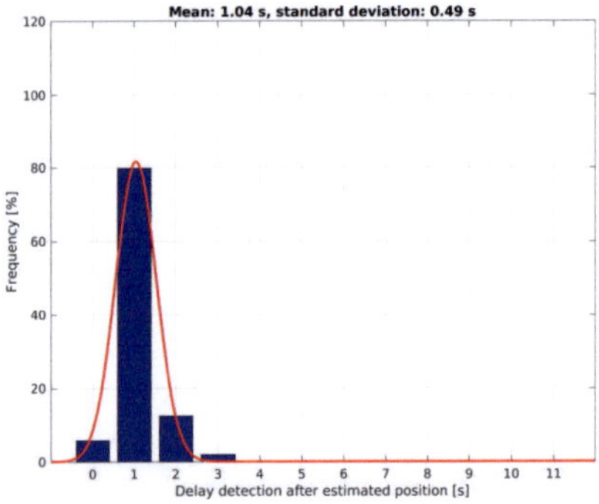

(c) Frog delay in trailing direction.

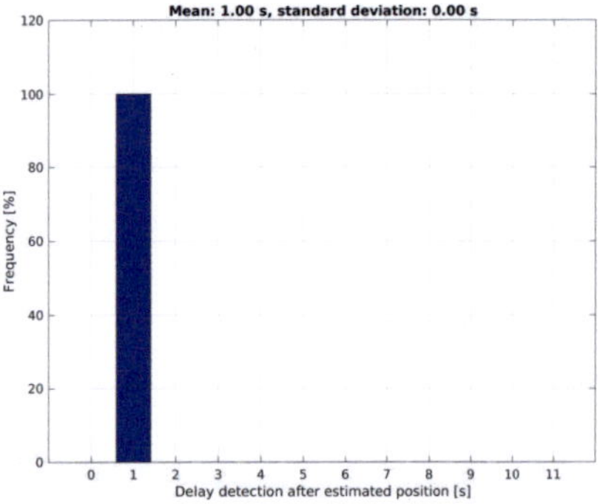

(d) Blade delay in trailing direction.

Figure 3.19: Temporal delay between the estimated reference position and their detection on single turnouts. The histograms have a bin size of 1 second. The distribution of Δt_{frog} and Δt_{blade} respectively is shown in blue, while the red curve represents — where appropriate — the corresponding probability distribution, whose mean and standard deviation are denoted above each figure.

Moreover, the shape of the histogram of the spatial distance between the tip of the frog and its detection in trailing direction (cp. Figure 3.18c) is quite similar to those in facing direction (cp. Figure 3.18a). Since the railway vehicle needs to pass the frog and additionally at least parts of the blade for a detection in trailing direction, there is another offset, although it is smaller. In contrast, the distribution of the spatial distance between the tip of the blade and its detection in trailing direction (cp. Figure 3.18d) looks quite arbitrary. Since the

tip of the blade is always detected 1 s after its passing in trailing direction (cp. Figure 3.19d; which equals the chosen sliding window size), Figure 3.18d only reflects the distribution of the distance traveled at the respective velocity of the railway vehicle within this interval.
Overall, the blade has been detected in 99 % of all cases within less than 10 s after passing its tip (cp. Figure 3.19b and Figure 3.19d), while the extended delays are exclusively caused by smallest velocities when starting from standstill or slowing down respectively. Moreover, the frog — in the considered data sets — has always been detected less than 3 s after passing its estimated tip (cp. Figure 3.19a and Figure 3.19c). Thus, each event is raised close in time to the detected tip of the frog or the blade. This can help to assess several hypotheses within a train-borne localization system, which, inter alia, is considered in the subsequent chapter.

Interim conclusion

Not only has almost every single turnout on the ego track been detected correctly, but also the branching direction thereon. Additionally, the deviation of the estimated position from its reference is in the smallest single-digit meter range for both the tip of the frog and the tip of the blade. Finally, in the majority of cases, the corresponding event has been raised within 10 s after the detected tip of the frog or the blade and often much earlier. Thus, in addition to the topological information, the proposed approach provides the direction of the respective branching maneuver and the corresponding reference position with high accuracy.

4 Discussion

In this chapter, the experimental results are analyzed and overall conclusions are drawn. Afterwards, the proposed approach and its experimental results are put into context with the state of the art. Thus, a classification of the contribution within a broad perspective can be achieved.

4.1 Review of the proposed approach

The experimental evaluation in Chapter 3 already showed promising results and proved that the proposed multistage approach is able to deal with complex situations. Although some imperfections resulted from the characteristics of the test ground and will thus most likely not occur in a common railway scenario, three main issues have been identified. Their impact is analyzed and potential countermeasures are proposed. Based on this, conclusions from the experimental results of the proposed approach are drawn. Furthermore, opportunities are shown in terms of how the detection of rails and tracks can be adapted to varied railway environments. Finally, this section demonstrates in which way the detection of both topology and branching directions on turnouts can promisingly be integrated into an application, such as a train-borne localization system.

4.1.1 Remaining limitations and their elimination

Based on the results presented in Chapter 3, three main causes for limitations have been identified. For each, the impact is analyzed and potential countermeasures are pointed out where necessary.

The first type of limitation occurs when a track is located partially or completely outside of the evaluation range as introduced in Section 2.3.1. Such situations particularly arise when the lateral distance between the ego and another (parallel) track exceeds 5 m. Since at least one rail is temporarily not "visible", track sections cannot be detected and the corresponding topology events are thus not raised. Its impact can be reduced by enlarging the evaluation range specifically in x-direction[36,37]. However, this limitation type and the proposed countermeasures only influence the completeness of track and topology detections. Neither do they affect the completeness, nor the correctness of the detection of turnouts on the ego track and the branching direction thereon. For turnouts on the ego track, the chosen evaluation range is already perfectly adequate.

The second limitation type is caused by extensive occlusions of the rail profile. Unlike the first limitation type, it affects both the ego and all other tracks. While buried rails are comparatively rare (and the thereby slightly decreased completeness of topology events is accepted subsequently), covered tracks can be found, inter alia, on level crossings. Since the rail profile is not entirely visible, the template matching approach introduced in Section 2.3.4 rejects the related measurements due to their low similarity with the assumed Vignole rail profile. Thus, track sections and the corresponding topology events cannot be detected within these areas. As long as level crossings occur only sparsely and turnouts on the ego track are not located within these areas, the resulting reduction in the completeness of topology events can be accepted. Otherwise, the approach can be extended by a model of their profile. Nonetheless, several types of construction exist which use Vignole as well as grooved rails[38], while the smaller visible dimensions make their detection even more difficult.

[36] However, there is an upper limit (cp. Section 2.3.1). For the considered setup described in Section 3.1.1, it is around ± 8 m in x-direction.

[37] An enlargement in z-direction enables the detection of further nearby tracks, which are located on a lower or higher level compared to the ego track. The conclusions drawn for the x-direction apply analogously for this.

[38] Their profile and main dimensions are for example defined in [EN 13674-1] and [EN 14811] respectively.

However, approaches for the detection of level crossings in general [Rahmig et al. 2013] and grooved rails in particular [Stein et al. 2016a] have been proposed. Thus, the completeness of the detections can be increased again, which applies for all steps starting with rail profiles and ending up with topology and branching directions.
Finally, the third limitation type results from two incidents which are analyzed separately at first. On the one hand, rails might as well not be visible. This happens in extended areas when the corresponding track is occupied by another railway vehicle or when noise barriers are located between the other track and the respective railway vehicle. The resulting missing track and topology detections are insignificant due to the fact that lidar sensors — as most of the other concepts — can neither penetrate those objects, nor look behind them. However, parts of the closer rail below the other railway vehicle might be detected. On the other hand, the assumption of one common rail profile might not be applicable everywhere[39] without it being a problem of visibility. The template matching approach at least tolerates slight deviations from the rail profile. Missing rail detections especially occur near the frog and are thus rather limited. Some erroneous detections caused by extended objects with rail like shape even remain after the detection of rail sections.
When the drawbacks of both incidents occur closely together, actually uncritical phenomena might result in erroneous track detections and corresponding topology events. Anyway, potential countermeasures exist on different stages. The side of a rail within a track (i.e., left or right in the direction of travel) might already be recognized in the profile detection and could be used for its further evaluation. However, the shape of Vignole rails is symmetric and thus

[39] While straight and curved tracks typically have the same rail profile [EN 13674-1], parts of turnouts (cp. Figure 2.24 and [EN 13232-2]) might require a special design, for example rails of the switch blades [EN 13232-5], the frog [EN 13232-6], or check rails [EN 13232-3]. Furthermore, diagonal tracks in particular lead to an affine transformation of the visible rail profile (strictly speaking, a rotation followed by a non-uniform scaling and a translation). In both cases, this leads to deviations from the ideal cross section profile.

its side is indistinguishable in such way. A better approach is the investigation of the orientation of all rails potentially forming a track as proposed for example in [Hackel et al. 2015][40]. This might increase the accuracy of track and topology detections. Yet, the optimal solution is the awareness of the potential inaccuracy for each event and its consideration in the overlying application, such as a train-borne localization, which is described exemplarily in Section 4.1.4.

In conclusion, the impact of many causes is limited. For all other cases, countermeasures have been proposed and partly already implemented.

4.1.2 Conclusions from the results

The focus on primary characteristics of the railway network, which are necessary for the running of the railway vehicle and are thus available in almost every scenario, allowed to receive the desired information:

- The proposed multistage approach makes use of the basic structure of a railway network (cp. also its overall structure in Figure 2.2). It thereby considers the most commonly used railway infrastructure elements: *(Vignole) rails and tracks.*
- It not only allows to recognize the ego track, but also works when surveying the nearby environment — irrespective of whether the tracks are parallel or diagonal therein. From that, the *topology of the railway network* can be derived.
- Moreover, it detects the position of that element, which makes the change from one track into another possible: *single turnout.*
- Additionally, it determines the chosen direction when driving across single turnouts: *branching direction thereon.*

[40] Information on the movement of the railway vehicle is required for this, so that 3d coordinates can be determined. Since the proposed approach solely considers 2d lidar sensor data, parallelism can only be checked heuristically as described for example in Section 2.4.1.

- Finally, these detections have become possible by using *only 2d lidar sensor data* without the help of any supplement, such as external illumination, information on the movement[41] of the railway vehicle, or location-based data from a digital map.

From the results in Chapter 3, it can be concluded that the proposed setup and the lidar sensor are capable of detecting the relevant information on the environment, while countermeasures for the few remaining limitations have already been proposed in the previous section.
The results are reliable and repeatable. Above all, this can be seen from the experiments on the circuit, which has been traversed in opposed driving direction and with different velocity profiles without a significant effect on the quantitative results (cp. Figure 3.10). Furthermore, demanding railway station scenarios with several single turnouts and many different branching situations did not influence their detection. The experiments on the siding and exchange tracks, which represent such scenarios, have proven this fact. There and in general, 98 % of the turnouts and the branching directions thereon have been detected. The same quantitative value has been reached for the accuracy of both the branching and topology events (cp. Table 3.3 and Table 3.4), which substantiates the strengths of the proposed approach. Moreover, the position of the tip of the blade as major reference position on a single turnout has been determined very accurately. This is evident from an overall mean of less than 10 cm and a standard deviation of about 1 m (cp. Figure 3.17).
Furthermore, the proposed approach performed well even in scenarios that provide tracks for meter and normal gauge. While the increased complexity of the corresponding single turnouts results in a few erroneous detections thereon, common railway networks do not require such complex infrastructure elements. Thus, the accuracy might be even higher there.

[41] Although a movement of the railway vehicle in the direction of travel is shown in most figures, this information is only considered for illustration purposes, but not in the proposed approach.

Finally, the detected events provide valuable information on the topology of the railway network and the branching maneuver, which the railway vehicle has performed therein. How the proposed approach can easily be adapted to varied environments is shown subsequently.

4.1.3 Adaptability to varied environments

Based on the promising results, it is shown in which way the proposed approach can be adapted to varied environments. This includes possible modifications resulting from the railway vehicle which hosts the corresponding setup on the one hand. On the other hand, different railway network characteristics might affect the multistage approach. It can be adapted to varied geometrical dimensions and extended to detect other connection types.

Railway vehicle-related modifications

The characteristics of the railway vehicle primarily influence the setup which has been proposed in Figure 2.3. The experimental railway vehicle, which has been used for the evaluation, allowed an upright mounting of the lidar sensor (cp. Figure 3.1). In general, the inclination angle should be as small as possible, since otherwise, vibrations or reflections might strongly disturb the measurements. However, tilting cannot always be prevented, which can be seen from the steep front of modern high-speed trains such as the ICE 4 [Siemens 2016][42]. Anyway, those inclined measurements can be transformed, so that they describe the shape in the x-z-plane again (cp. Section 2.2.1).
Independent of the design of the front, the lidar sensor should be mounted in the highest possible position (cp. Section 2.2.1). In contrast, an installation below the railway vehicle leads to an increased proportion of occluded rails, which

[42] A strongly inclined mounting decreases the share of the laser beam that is reflected to the lidar sensor, which might result in information loss. Additionally, the bigger the tilting is, the higher is the leverage effect of vibrations, which adversely affects the measurement results.

form the basis for all detections[43]. Missing information cannot be recovered, so that mounting too low is not recommendable.

The measurement rate of the lidar sensor should conform to the maximal velocity of the railway vehicle. As soon as the railway vehicle runs with double speed, the distance in driving direction between consecutive scans increases accordingly. Too little information on the environment can be obtained in this way and thus, the detection of rail and track sections might be prevented. However, this effect can be compensated by an increased measurement rate of the lidar sensor[44].

In summary, the proposed setup can easily be adapted to varied characteristics of the railway vehicle. In addition to the described modifications, the current admission requirements need to be considered for an operation in a railway scenario[45].

Adaptations to other railway network properties

The characteristics of the railway network primarily influence the approach proposed in Chapter 2. Possible adaptations resulting from varied environments are discussed accordingly. Finally, it is shown how branching directions on further connection types might be detected when using the proposed approach for single turnouts again.

[43] Occlusions make the detection of nearby tracks and tracks branching from or into the ego track difficult or even impossible. Thus, they reduce the completeness of the topology events and particularly prevent the detection of turnouts and the branching direction thereon.

[44] For high-speed trains, a measurement rate in the lower three-digit range is required, while the angular resolution can remain unchanged compared to the experimental setup (cp. Section 3.1.1). However, such lidar sensors currently cost a six-figure EUR sum. Furthermore, the resolution in y-direction also depends on the requirements of the specific application.

[45] While [Stein et al. 2016a] derive functional demands on the measurements of a lidar sensor, such commercial off-the-shelf products have to be further toughened up for their permanent use in a railway scenario. For example, the lidar sensor needs to be electromagnetically compatible [EN 50121], have a proven robustness against varied climatic and operating conditions [EN 13848-2], and meet the needs of reliability and safety [EN 50126]. However, such a customization is beyond the scope of this work.

The discussion of the limitations already showed that the detection does not rely on a perfect match between the considered rail profile and its measurements. However, the used template library can be prepared for every Vignole rail shape (cp. also their typical dimensions in Figure 2.5 and Table 2.1). At the same time, the gauge, which is considered in the detection of track sections, can be adapted to the value that is used in the according railway scenario. Finally, the detection of track sections can even be extended to detect every mixed gauge scenario, which has already been demonstrated by the detection of both gauges on the test ground (cp. at the end of Section 2.4.2).

Even the use of an explicit geometrical track model might be possible. However, it would not prevent the few missing detections of track sections. On the one hand, tracks that are almost perpendicular to the ego track will hardly be ever detected, since the profile of their rails cannot be observed within a scan. On the other hand, some parts of turnouts do not have a parallel counterpart, which results from their function. This holds in particular for parts of the blade, but also for wing and check[46] rails where no parallel rail with a lateral distance of at least the minimal gauge can be found. In fact, the heuristic detection of track sections proposed in Section 2.4.2, which considers only the minimal allowable gauge, even enables an association of rail sections which are not yet perfectly parallel[47]. Thus, it also tolerates slight deviations in the lateral position, which might result from erroneous measurements or an imprecision in the execution of previous steps.

In addition, the proposed approach can be extended to further connection types. Since complex turnouts are effectively assembled from interleaving single turnouts, the detection of single turnouts from Section 2.5.2 can be utilized.

[46] The influence of check rails on the detection results should be further examined experimentally. Since none of the passed turnouts contained them, it can just be assumed, that their impact is comparable to those of wing rails. The detection of track sections has been slightly delayed, but they did especially neither influence the completeness nor the accuracy of branching events.

[47] A further check on the parallelism of rails or tracks would again require 3d information, which is not used herein (cp. also footnote 40 on page 96).

For this purpose, it is possible to apply the concept of branching directions to less frequent connection types (cp. Figure 4.1). Again, it is irrelevant whether tracks are straight or curved. Two structurally different cases have to be distinguished: three-way switches and connections based on diamond crossings.

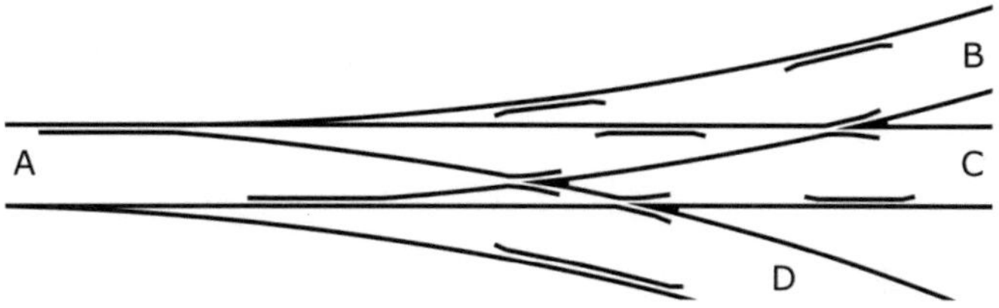

(a) Three-way switch.

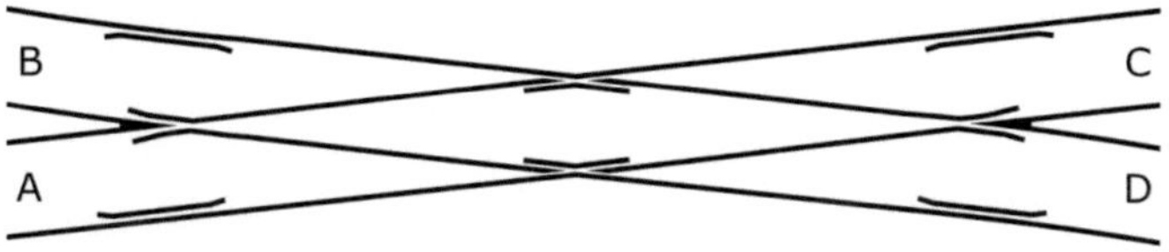

(b) Diamond crossing.

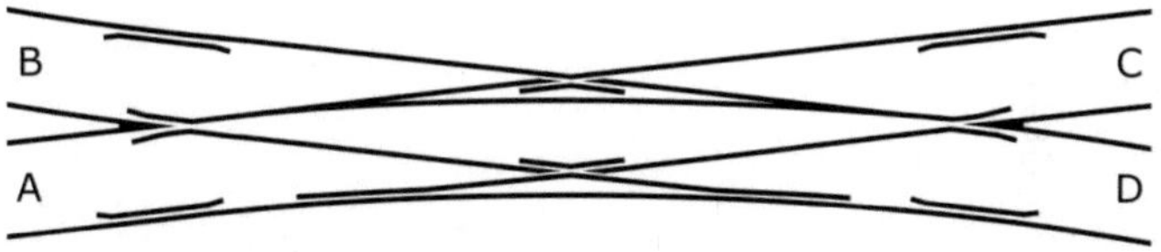

(c) Diamond crossing with single slips.

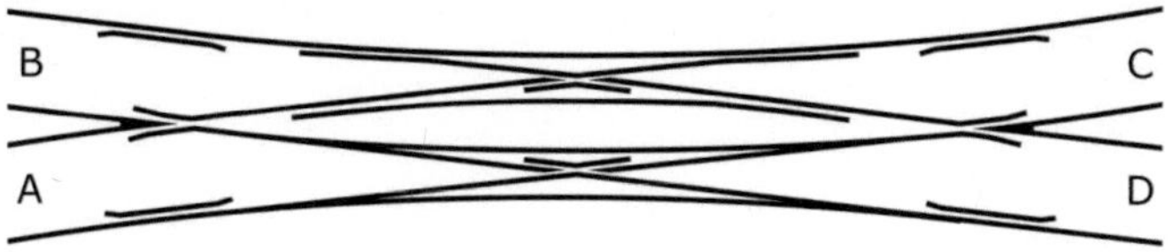

(d) Diamond crossing with double slips.

Figure 4.1: Complex connection types. A to D denote positions thereon.

In the first case, three-way switches are considered. They are assembled by two interleaving single turnouts. When driving in reading direction in Figure 4.1a, the first single turnout allows to drive either to position D or to the second single turnout, which heads either for position B or for C. All expected detections of branching directions are summarized in Table 4.1[48].

Table 4.1: Branching directions when passing complex connection types, such as three-way switch, diamond crossing, and diamond crossing with single or double slips as outlined in Figure 4.1. The already considered passing of a single turnout is additionally listed (cp. also Figure 2.27). A to D denote positions thereon, while FL ... TR represent the expected branching direction (facing left ... trailing right). If a branching maneuver is not possible, the corresponding direction is represented by ÷, whereas route C→D and vice versa occurs in no case.

	branching direction(s) on				
			diamond crossing		
route	**single turnout**	**three-way switch**	**"pure"**	**single slips**	**double slips**
A→B	FL	FL FL	÷	÷	÷
A→C	FR	FL FR		FL	FL TL
A→D	÷	FR	÷	FR TL	FR TL
B→A	TL	TL TL	÷	÷	÷
B→C	÷	÷	÷	÷	FL TR
B→D	÷	÷		TR	FR TR
C→A	TR	TR TL		TL	FL TL
C→B	÷	÷	÷	÷	FR TL
D→A	÷	TR	÷	FL TR	FL TR
D→B	÷	÷		FR	FR TR

The second case comprises three different connection types: diamond crossings (without a name affix) as well as diamond crossings with single or double slips. Diamond crossings are an intersection of two tracks which provide four routes in total (A→C, B→D, and vice versa; cp. the positions in Figure 4.1b).

[48] Note that in two cases (A→D and vice versa) only one single turnout is passed, so that only one branching direction detection is expected thereon.

Since no single turnout is used, no branching is possible and thus no corresponding detection is expected thereon (cp. Table 4.1). Diamond crossings with *single* slips are assembled from a diamond crossing and two interleaving single turnouts. Besides the four routes on the diamond crossing, they allow an additional diagonal passing from position A to D (and vice versa as shown in Figure 4.1c). While those diagonal maneuvers cause two detections of branching directions, a single detection is expected for a straight route (cp. Table 4.1). Finally, two further interleaving single turnouts allow another diagonal passing from position B to C (and vice versa) on diamond crossings with *double* slips (cp. Figure 4.1d). Since each of the eight potential routes passes two single turnouts, two branching directions can be expected in each case (cp. Table 4.1)[49].

Thus, the branching direction detection on complex connection types reduces to the detection of the interleaving single turnouts therein. However, this has already been solved, as extensively discussed before. Even if those further connection types cannot (correctly) be identified[50], though the corresponding tracks have most probably already been detected in the topology events (for example when driving from position A to C as shown in Figure 4.1a, topology events on the left and right side for the track towards position B and D respectively can be expected).

The detection of turnouts beyond the ego track is conceivable and has been investigated for example in [Ponciano et al. 2015]. However, due to the occlusions

[49] The first passing always occurs in facing direction, while the second one is in trailing direction. The corresponding side (left or right) is equal for straight passings, whereas it is different for diagonal ones.

[50] An additional detection of the connection type is conceivable. However, complex turnouts are rare compared to single turnouts. Since the test ground does not contain any complex connection type, the detection of them or the branching direction thereon cannot be evaluated. It can simply be assumed that, due to their increased complexity, the detection of the branching direction is more difficult compared to those on single turnouts. Otherwise, lateral distances between rails, which are smaller than the considered gauge, can be an additional indication for their existence, as done in [Hackel et al. 2015; Oude Elberink and Khoshelham 2015]. However, valid statements on the correctness and completeness of their detections and the accuracy of the estimated reference position require corresponding experiments.

of their rails, this is much more difficult compared to turnouts on the ego track and might result in a limited completeness. Nonetheless, even if those turnouts had been detected, the information on the corresponding "branching direction" would not be useful for the railway vehicle on the ego track.
In summary, the proposed approach can easily be adapted to varied characteristics of the railway network. It can even be extended to further connection types which for example can be found in railway station scenarios where many tracks are interconnected closely together. Furthermore, it is generic in so far as every complex turnout is assembled from interleaving single turnouts, whose detection has already been evaluated experimentally. How an exemplary application can utilize the additional information from the herein proposed detections, is outlined subsequently.

4.1.4 Integration into a self-localization framework

Finally, train-borne localization is examined as an application. At first, the major challenge of those system is identified. Furthermore, it is shown how the resulting ambiguities and uncertainties are considered therein. For their elimination, the branching and topology events proposed within this work might provide an added value. Thus, requirements for their integration into such a self-localization framework are identified. Finally, it is demonstrated that these requirements are already met by a prototypical implementation.
In the following, a digital map of the railway network is assumed as given. It describes the relevant geometrical and geographical features of the railway network and its topology, where several connection types are considered[51]. How-

[51] Potential data sources are railML [Nash et al. 2004] or OpenStreetMap [Rahmig and Kluge 2013]. In each case, the herein considered framework and its continued development provide appropriate parser. They ensure that the relevant information on the infrastructure of the railway network can be extracted, so that the required topographical map is created.

ever, the information from topology and branching events can at least be used for a topological mapping of the nearby environment[52].

Characteristics of current self-localization approaches

State of the art localization systems which use longitudinal and geographical position information and a digital map are not always track-selective (cp. Section 1.2). For these, it is especially difficult to determine on which of two or more parallel tracks the railway vehicle is located. Self-localization approaches, such as [Lauer and Stein 2015], consider three main issues in order to avoid erroneous decisions and thus enable a safe operation:

- Instead of promptly deciding on the most likely, but possibly erroneous location of the railway vehicle, such approaches explicitly consider ambiguities of potential positions in the railway network and their individual probability (Bayesian multi-hypothesis approach).
- Each measurement of the sensors is regarded as inaccurate. Thus, not only its value, but also uncertainties are considered (stochastic modeling in terms of distributions).
- The corresponding method decides on the generation and elimination of hypotheses[53] depending on the topological situation and the probability of the hypotheses. Furthermore, it considers those uncertainties in derived quantities and even allows to integrate delayed measurements (incremental processing and prediction).

[52] If information on the movement of the railway vehicle is directly considered, so that 3d coordinates are available, a topographical map could be derived.

[53] A hypothesis corresponds to a potential position of the railway vehicle. In [Lauer and Stein 2015] for example, they are described by a quadruple (track, orientation thereon, longitudinal position along this track, and velocity of the railway vehicle) in the form of a Gaussian mixture distribution and their individual probability.

Capabilities of the proposed events

Ambiguous situations in particular arise when (single) turnouts are passed in facing direction. In this case, the previous position needs to be split into one hypothesis per connected track, whereby each track is driven on with the same probability at the moment.
However, the mandatory part of topology and branching events described in Table 2.2 and Table 2.4 respectively already provides valuable information on the environment and especially the route driven therein. The estimated tip of the blade[54] on a single turnout represents a specific longitudinal position. Furthermore, the detected branching direction thereon allows to distinguish the lateral position: in terms of the passing direction (facing or trailing), the side of the chosen track (left or right), and thus the probability of the corresponding hypothesis. At least the rating of hypotheses can be changed in this way. Hypotheses which comply with the detected branching direction become more probable, while all others are less likely. Further observations of neighboring tracks (tracks detected on the left and/or right side at a specific longitudinal position) might lead to the exclusion of unlikely positions.

Requirements for the integration of events into a self-localization approach and their implementation

Four requirements have been identified as prerequisite for the integration of the proposed events into a self-localization approach: incompleteness, inaccuracy, deviations, and past events. They base on the characteristics of the probabilistic localization framework and the features of topology and branching events. In the following, each requirement is described and the implementation approach, which is used in the prototypical integration, is given. Additionally, each requirement is illustrated using the example of passing a single turnout.

[54] The tip of the blade has been identified as the more relevant reference position compared to the also determined tip of the frog in Section 2.5.2 and the experimental results.

1. **Absolute completeness of topology and branching events cannot be guaranteed:**
   - *Description:* The occurrence of events strongly depends on the spatial situation. On the one hand, branching events only occur at certain places, namely on turnouts, and are expected at most once thereon. On the other hand, topology events can be raised several times and almost everywhere, as long as neighboring tracks are in the nearby environment. However, occlusions by noise barriers or other railway vehicles might result in missing detections.
   - *Implementation:* Since no general rule on the occurrence of those events can be derived, their absence is implicitly considered in the method. Inspired by [Johannes and Almeida 2014], only "positive" observations (regardless of whether they might be evaluated as a true or false positive afterwards) can have an influence on the position and probability of hypotheses, while missing observations must not.
   - *Example:* A missing single turnout detection does not have an impact on any hypothesis. Furthermore, it is not concluded on their absence, so that the previous state is kept.
2. **Total accuracy of topology and branching events cannot be assumed:**
   - *Description:* Each event can be erroneous, which results for example in "phantom tracks" or in repeated detections of single turnouts with different gauges.
   - *Implementation:* The incorrectness of each event is explicitly considered, as proposed by [Rahmig et al. 2013]. Unlike the previous case, their overall error probability can be determined experimentally. The precision quantifies the accuracy of the corresponding event, which is denoted in Table 3.3 and Table 3.4 for topology and branching events respectively.

- *Example:* A single turnout is passed in facing right direction and this branching direction has been detected. In the following, the probability of the corresponding hypothesis increases, while for example a potential hypothesis driving in facing left direction becomes less likely, but it is not excluded, since the detected branching direction might be erroneous (cp. the size and saturation of the triangles in Figure 4.2b).

3. **Deviations have to be taken into account:**
    - *Description:* Similar to other measurements, the longitudinal position of the event does not have a fixed value, but might deviate from it.
    - *Implementation:* While the side of a topology event and the detected branching direction are categorical data, only the position uncertainty of the estimated tip of the blade needs to be considered explicitly[55]. It is yet again modeled as Gaussian, whereas its distribution depends on whether the single turnout is driven on in facing or trailing direction. The mean and the standard deviation from Figure 3.17b and Figure 3.17d denote the corresponding values respectively.
    - *Example:* When passing a single turnout, the longitudinal position of the hypotheses might be aligned towards the estimated longitudinal position of the blade[56], which also allows to compensate a drift caused by other sensors.

[55] Unlike many other approaches, [Lauer and Stein 2015; Pottberg 2016] explicitly consider the uncertainty of the given digital map in addition to those of the sensor measurements and the estimated reference positions of the events (cp. the search areas in Figure 4.2). In contrast, [Rahmig et al. 2013] for example only consider the standard deviation of the detected turnout position. While the examination of the topological correctness of the digital map is relatively simple, the verification of its geometrical and geographical accuracy would result in great effort for reference measurements, since an up-to-date information is usually not available.

[56] Additionally, it might be considered that the detection of the turnout itself is erroneous. For this purpose, another hypothesis needs to be established. This keeps the original longitudinal position, but its probability will be reduced significantly.

4. **Detections might be located in the past**:
   - *Description:* Events do not solely base on the current lidar sensor measurements, but rely on their progression as well.
   - *Implementation:* The method already explicitly considers the fact that measurements might be delayed and need to be integrated subsequently. This holds in particular for branching events, since the position of the tip of the blade is estimated after its passing. This delay results from the dimensions of the turnout, the velocity of the railway vehicle, and the direction of the passing. Corresponding quantitative values are shown in Figure 3.19b and Figure 3.19d for a passing in facing and trailing direction respectively.
   - *Example:* When passing a single turnout in trailing direction, the blade is promptly detected. A passing in facing direction results in a higher delay between the tip of the blade and its detection due to the larger extension of the single turnout. Thus, the method needs to wait longer, whereby at least a common threshold should be considered therein.

Based on this work and the described requirements, topology and branching events have already been implemented prototypically into a probabilistic localization framework (cp. [Pottberg 2016]). The corresponding visualization of exemplary scenarios in Figure 4.2 shows several hypotheses and the search area for topology and branching events in the digital map. These events affect the determination of the hypothesis positions and probabilities as illustrated in Table 4.2. The mathematical derivation of the probabilistic modeling of those events and their implemented integration into the probabilistic localization framework is described in [Pottberg 2016][57].

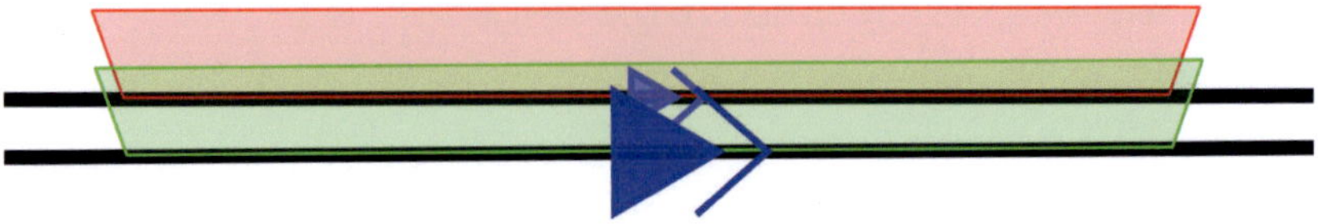

(a) Track on the left side: If the topology event supports the hypothesis, it is highlighted in green, while red represents the fact that no neighboring track is present in the digital map. Trapezes border the search area for tracks.

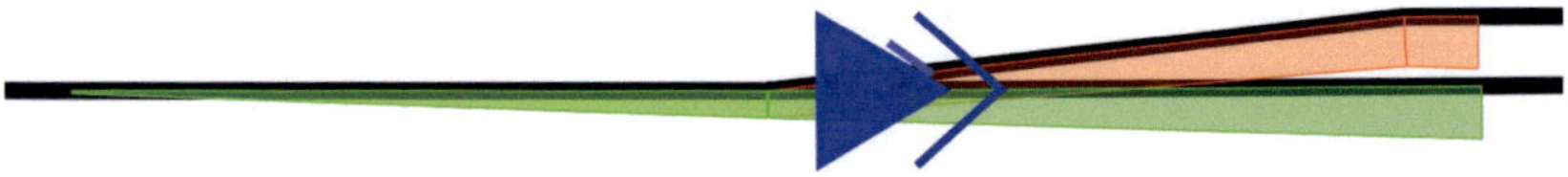

(b) Passing a single turnout in facing right direction: If the branching event supports the hypothesis, it is shown in green, while orange represents an inaccurate branching direction. The extension of the green or orange triangles along the track denotes the search area for turnouts.

Figure 4.2: Examples for topology and branching events in the implemented approach. Black lines represent tracks. Every triangle shows the position of a hypothesis, whereas its size and saturation reflect the corresponding probability.

Interim conclusion

Topology and branching events enable an additional and independent evaluation of the probability of the hypotheses and are thus useful to improve track-selectivity. In addition, the detected tip of the blade enables an increased (longi-

[57] It should further be noted that [Pottberg 2016] outperforms the concepts which have been published before. [Rahmig et al. 2013] only use the beginning and the end of level crossings for an adjustment of the longitudinal position of the railway vehicle, but cannot determine the position of a single turnout accurately. However, [Pottberg 2016] presents a method how the herein determined reference position is actually used for this. In addition, it considers the detected branching direction thereon and uses detected neighboring tracks for the evaluation of the probability of hypotheses, while [Johannes and Almeida 2014] just express these ideas. Moreover, [Pottberg 2016] also considers the incorrectness of each detection and quantifies the added value when using both events in the self-localization approach, whereas [Rahmig et al. 2013] only speculate on their advantages.

tudinal) position accuracy. Since only positive detections are considered[58], the individual benefit of each event depends on the topological situation. However, in a single-track scenario without any turnouts the major problem of current train-borne localization does not appear. First investigations in [Pottberg 2016], that also base on experiments and results of this work, have demonstrated that track-selectivity has been faster and in more cases restored when using those events compared to a self-localization that is solely based on GNSS and velocity measurements[59].

Table 4.2: Influence of the events on the position of a hypothesis and its probability. Each table entry denotes direct impacts and the thereby considered deviation.

event type	**affected feature of the hypothesis**	
	longitudinal position	probability
topology	no influence	topology precision
branching	blade position uncertainty	branching direction precision

4.2 Review of the state of the art

In recent years, the number of lidar sensor systems for the monitoring and mapping of the railway environment has increased [Beger et al. 2011; Puente et al. 2013; Reiterer et al. 2014; Mikrut et al. 2016]. Typical applications are the inspection or inventory of tracks (including the monitoring of the rail profile, the measurement of gauge, superelevation, or radii, and the capturing of

[58] The information about the absence of nearby tracks or turnouts on the ego track can also be modeled in the self-localization framework, which requires their reliable detection. However, it is extremely difficult to distinguish whether a nearby track is for example covered by snow, or occluded by a noise barrier, or whether it is actually not present.

[59] While no erroneous branching event occurred in [Pottberg 2016], nine false positive topology events had no apparent impact and did especially not lead to wrong position estimates.

the clearance gauge) [Leslar et al. 2010; Soni et al. 2014; Mikrut et al. 2016], tunnels [Kang et al. 2014], or overhead contact lines [Jung et al. 2016]. Stationary, driving, as well as flying systems are used therefore which are referred to as terrestrial, mobile, and airborne laser scanning systems (abbreviated as TLS, MLS, and ALS respectively). While point density and measurement accuracy typically decrease from TLS to ALS, the covered area increases in return. However, specialized software to process and interpret the data is often needed [Neubert et al. 2008; Leslar et al. 2010; Kremer and Grimm 2012; Soni et al. 2014] or data has to be selected manually [Mohamad et al. 2013], which is almost exclusively possible in post-processing. Since an online application such as train-borne localization cannot wait for information that is generated after the completion of the journey, but close to the measurement time, a causal and real-time capable approach is required.

No common approach exists with regard to MLS and the detection of relevant infrastructure in this data. Moreover, the number of publications covering their *individual* method at least for parts of the considered rails, tracks, and turnouts is modest. Table 4.3 gives a comprehensive overview[60], while strengths and weaknesses of related work are addressed subsequently.

Not included therein is [Daoust et al. 2016], which only rely on a horizontally scanning lidar sensor, but do not capture rails. It allows to localize the subway in tunnels[61] and relies on the homogeneous shape of the environment, whereas common railway scenarios change dynamically (e.g., vegetation depending on the season) or such nearby objects are not present.

[60] For example, [Arastounia 2015] primarily focuses on overhead contact lines, but also detects rails for this. [Neubert et al. 2008] use ALS data, but projects them on cross-sections along the track, which in turn resembles MLS measurements. Thus, these publications are also included.

[61] Previous lidar sensor measurements of each track are required for this. Their extent increases with the size of the railway network, whereas this work only needs the appearance of a rail profile in a single scan.

4.2.1 Setup, sensor, and data

Within the related work, the system is often mounted on the front of the railway vehicle or on its top. Especially the tilted setup in [Rahmig et al. 2013; Johannes and Almeida 2014] allows the detection of objects, such as turnouts, before they are passed by the railway vehicle. Additionally, the inclined measurement of rails results in increased height differences and might thus ease their detection. However, the risk of reflections and the disturbance by vibrations increase as well (cp. Section 2.2.1).

The lidar sensors used in the related work can be divided into two categories, which differ in performance and acquisition cost. The first one includes highly specialized mapping systems that often consists of two lidar sensors. Since their total number is limited, those are very expensive. In contrast, [Rahmig et al. 2013; Johannes and Almeida 2014] and this work use low cost lidar sensors with adequate accuracy[62].

While almost all approaches of the first category require 3d information at least for parts of their subsequent processing, the methods of the second category benefit from the known arrangement of the measurements within a scan. However, vertical cross-sections are also directly available due to the upright mounting of the lidar sensor in [Blug et al. 2004; Ponciano et al. 2015; Hackel et al. 2015] or indirectly by projecting measurements on vertical planes in [Neubert et al. 2008; Yang and Fang 2014].

In summary, all MLS approaches provide measurements of at least the herein considered detection area, which covers the ego track and nearby tracks, while the lateral and longitudinal resolution of the data differs.

[62] The measurement accuracy is — with regard to the present state — about 0.5 cm for the first category and between 1 and 3 cm for the second one (such as this work). The angular resolution ranges between $1/6^\circ$ and $1/14^\circ$, while the measurement rate ranges from 25 scans/s [Johannes and Almeida 2014] to about 300 scans/s per lidar beam for the chosen angular resolution (this work: $1/14^\circ$ and 50 scans/s). All systems at least allow measurements in a semicircular area up to 10 m, which is sufficient for the detection of the ego track as well as nearby tracks. However, there is a large price range starting about 6,000 EUR (this work) and significantly exceeding 100,000 EUR.

Table 4.3: Overview of related works in rail, track, and turnout detection from lidar sensor data for driving vehicles. If an approach meets a criterion wholly or partially, it is denoted by checks ✓and ✓ respectively.

approach	additional input: explicit geometrical model	additional input: echo information	output: rails	output: neighboring tracks	output: turnouts	output: branching directions	quantitative results
this work[a]	✓	✓	✓	✓	✓	✓	✓
[Blug et al. 2004]		[b]	✓				
[Neubert et al. 2008]	✓		✓	✓[c]			
[Rahmig and Johannes 2013]	✓	[d]	✓[e]	✓[f]	✓[e]	✓[e]	
[Rahmig et al. 2013]		[d]	✓	✓[f]	✓[e]	✓[e]	
[Oude Elberink et al. 2013]	✓		✓	✓[c]			✓
[Johannes and Almeida 2014]			✓	✓[f]	✓[e]	✓[e]	
[Yang and Fang 2014]		✓	✓	✓			✓
[Oude Elberink and Khoshelham 2015]	✓		✓	✓	✓[g]		✓
[Hackel et al. 2015]	✓		✓	✓	✓		✓

[a] Includes [Stein et al. 2016b].
[b] States characteristics of echo near rails, but does not use it for their detection.
[c] Not presented, but expected.
[d] Only used for the detection of level crossings.
[e] Vaguely described.
[f] Only in proximity to ego track.
[g] Even on neighboring tracks, but not in parallel with geometrical model.

Continued Table 4.3: Overview of related works in rail, track, and turnout detection from lidar sensor data for driving vehicles. If an approach meets a criterion wholly or partially, it is denoted by checks ✓ and ✓ respectively.

	additional input		**output**				
approach	**explicit geometrical model**	**echo information**	**rails**	**neighboring tracks**	**turnouts**	**branching directions**	**quantitative results**
this work[a]	✓	✓	✓	✓	✓	✓	✓
[Jwa and Sohn 2015]			✓				✓
[Arastounia 2015]			✓				✓
[Ponciano et al. 2015]	✓		✓	✓	✓[h]		✓
[Arastounia and Oude Elberink 2016]	✓		✓	✓			✓
[Arastounia 2017]			✓	✓			✓

[a] Includes [Stein et al. 2016b].
[h] Even on neighboring tracks.

4.2.2 Method

The related work differs with regard to the used approaches, the level of detail in the modeling, and the detectable objects. The first part of this section considers rails and tracks, whereas the second one focuses on branching situations.

Detection of rails and tracks

Several approaches for the preprocessing and detection of rails and tracks have been proposed. They differ in the individual procedure as well as in the utilized detail of railway knowledge and lidar sensor information, while all have in common that they at least detect rails (cp. Table 4.3).
Various methods have been proposed to restrict the search space. Potential points on the track bed are often detected at first, which can be excluded subsequently as done in [Oude Elberink et al. 2013; Yang and Fang 2014; Oude Elberink and Khoshelham 2015; Arastounia 2015; Arastounia and Oude Elberink 2016; Arastounia 2017]. The processed 3d data can also be disassembled in stripes with a fixed longitudinal size [Jwa and Sohn 2015].
Furthermore, several assumptions are made. Almost generic examples are the fact that rails have a continuous course with smooth curvature or that tracks are formed of pairs of rails with a similar alignment and a certain lateral distance. Even the assumption of (piecewise) straight courses or the fact that most points on rails are typically above the ground can often be fulfilled. Nonetheless, the expectation of at most one rail per meter in lateral direction [Oude Elberink et al. 2013; Arastounia 2015] disregards turnouts and diamond crossings. Besides, the assumption of an overhead contact line for each track [Arastounia and Oude Elberink 2016; Arastounia 2017] is too restrictive, since it would exclude for example almost one half of the German railway network[63].
The first proposed approach has been found in [Blug et al. 2004]. It detects the rails of the ego track within *fixed* areas of several accumulated vertical cross-sections. However, the detection and their check for plausibility solely bases on features such as relative positions within a scan, lateral distances, and differences in height. Since those detections form the basis for the inspection of the clearance gauge, no other tracks are considered.

[63] Based on statistics [European Commission 2014].

In [Neubert et al. 2008], a correlation based approach using a *simplified* geometrical model of the track cross-section has been proposed. However, since no turnouts are considered, problems especially arise in their environment. Further approaches subsequently made use of explicit geometrical models with different level of detail in 2d and 3d. They cover simplified geometries of cross-sections of rails [Rahmig and Johannes 2013; Hackel et al. 2015] or tracks [Ponciano et al. 2015] as well as straight sections of rails [Oude Elberink et al. 2013] or tracks [Oude Elberink and Khoshelham 2015], while this approach uses a detailed rail profile [Stein et al. 2016b]. Slightly different is the model used in [Arastounia and Oude Elberink 2016], which is a simplified top view on a straight track section.

In contrast, many approaches waive the use of explicit models and heavily rely on heuristics and descriptive statistics (e.g., quantiles and thresholds in histograms for the assignment of points to the ground [Oude Elberink et al. 2013; Oude Elberink and Khoshelham 2015; Arastounia 2017] or to rails [Arastounia 2015; Arastounia and Oude Elberink 2016]) and partially seem to be tailored to the considered example data. Furthermore, only [Yang and Fang 2014] consider the echo as additional information for the detection besides this work (cp. also *b* and *d* in Table 4.3), while all approaches use lidar distance measurements.

Rail and track sections are typically identified in multiple stages by fitting lines (e.g., RANSAC [Oude Elberink et al. 2013; Oude Elberink and Khoshelham 2015], least squares, and other optimizations [Hackel et al. 2015]), using clustering or region growing approaches [Oude Elberink and Khoshelham 2015; Ponciano et al. 2015; Hackel et al. 2015; Arastounia 2017], and different kinds of template matching (e.g., by correlation [Arastounia and Oude Elberink 2016] or even by estimating all six degrees of freedom for a given 3d model [Oude Elberink et al. 2013; Oude Elberink and Khoshelham 2015]). Remarkably, in [Oude Elberink and Khoshelham 2015] even the third adjacent track has been detected. However, the radius search used in many cases for almost every point is computationally intensive [Arastounia and Oude Elberink 2016; Arastounia 2017]. The same holds for the estimation of all six degrees of freedom, whereby these results need to be confined even further by a line estimation in [Oude Elberink et al. 2013; Oude Elberink and Khoshelham 2015]. Finally, the only detection of the ego track in [Blug et al. 2004; Jwa and Sohn 2015; Arastounia 2015][64] is insufficient for the purpose of this work, since no topology events can be raised (cp. also *f* in Table 4.3).

In summary, all approaches detect the rails of the ego track, while most of them also detect nearby tracks, so that topology events could be raised. However, quite different approaches and levels of detail are used for this, whereas some assumptions are too restrictive.

Detection of turnouts and the branching direction

While various approaches detect tracks, many of them ignore branching situations. Some authors also observed that their non-consideration results in problems in the rail and track detection and thus poorer results [Yang and Fang 2014; Oude Elberink and Khoshelham 2015].

For the detection, [Ponciano et al. 2015] use a "fixed logic with thresholds", which assembles X- and Y-shaped rail section patterns to turnouts and even classifies complex connection types. It is remarkable that [Oude Elberink and Khoshelham 2015; Ponciano et al. 2015] even detect turnouts on neighboring tracks, which are not relevant for the purpose of this work, though. However, neither a reference position, nor a branching direction are derived. The same holds for [Oude Elberink and Khoshelham 2015; Hackel et al. 2015], who detect turnouts by comparing the lateral distance between several nearby rails, their orientation, and arrangement, for which 3d information is required. [Rahmig and Johannes 2013; Rahmig et al. 2013; Johannes and Almeida 2014] already observed that particular elements on a turnout might be detected, such as the frog and check or wing rails. Furthermore, the lateral distance between the corresponding tracks increases from almost zero to approximately the gauge at the

[64] Note that especially the data set in [Jwa and Sohn 2015] contains four parallel tracks, but only one is considered.

tip of the frog when passing in facing direction. Finally, [Johannes and Almeida 2014] expressed the idea to detect the four herein considered branching directions from their chronological order. Yet, they were not able to determine a reference position on the turnout accurately, which might be used for a longitudinal position correction (cp. Section 4.1.4). It is furthermore striking that especially [Rahmig et al. 2013; Johannes and Almeida 2014] indicate promising ideas, but do not provide a specific method and the corresponding (quantitative) results (cp. also *e* in Table 4.3).

In summary, one half of the related work neglects turnouts, while some consider them. However, specific reference positions are hardly determined, whereas the approach has not been elaborated for the detection of branching directions.

4.2.3 Results and evaluation

A direct comparison of the related work is not feasible, since there is no common data set or benchmark. However, a part of the related work waives a quantitative evaluation (cp. Table 4.3). For example, [Neubert et al. 2008] only provide a visual inspection. [Rahmig et al. 2013] characterize their approach as "very stable for the detection", but just present one example.

If an evaluation is provided, most of the time only one data set with a length of a few 100 m is used that contains either a single track or several parallel tracks without branching, so that demanding branching situations cannot arise. In contrast, this work considered various scenarios and more than 200 branching situations. Finally, although [Arastounia and Oude Elberink 2016] emphasize their advantages compared to their previous works [Oude Elberink and Khoshelham 2015; Arastounia 2015], they neither used the same data set nor compared their approaches quantitatively.

With regard to quantitative measures, [Oude Elberink et al. 2013; Oude Elberink and Khoshelham 2015] provide deviation measures, but no statements on correctness and completeness. [Yang and Fang 2014] accomplish accurate and almost complete detections, but regard deviations up to 15 cm as properly, which

corresponds to twice the width of the rail head. On top of that, [Arastounia 2015; Arastounia and Oude Elberink 2016; Arastounia 2017] achieve almost complete and quite accurate rail detections where erroneous detections result from nearby vegetation or other raised objects. In contrast, [Jwa and Sohn 2015] detect only 82 % of the ego track. Remarkable is, that [Oude Elberink and Khoshelham 2015; Hackel et al. 2015] additionally use an independent reference system for their evaluation. The latter one reaches a mean deviation of less than 5 mm on the ego track. Furthermore, 35 out of 36 turnouts have been detected correctly, while three erroneous detections occurred. Finally, [Ponciano et al. 2015] detected 147 out of 151 turnouts, whereas missing detections resulted primarily from missing data and low point density on nearby tracks.
In summary, the related work often uses a single and short, but less complex data set for the evaluation and partially waives quantitative results. A few approaches achieve detection results comparable to the herein obtained ones. However, quantitative detection results of reference elements on single turnouts or the branching direction thereon are provided in no case.

4.2.4 Interim conclusion

This work provides a comprehensive system for the detection of all rail-related parts of the environment. It starts from rails, includes tracks and turnouts, and ends up with the branching direction thereon, while the number of related works in that order drastically decreases (cp. Table 4.3). It makes use of all information from a low cost lidar sensor, such as both distance and echo information and the temporal and spatial arrangement of the measurements. The most detailed model has been used, which still requires little memory. Finally and in contrast to most related work, it provides an extensive qualitative and quantitative evaluation and — along with [Lauer and Stein 2015; Pottberg 2016] — demonstrated to reduce ambiguities in train-borne localization.

5 Summary, conclusions, and future work

Herein, the main parts of the work are summarized. Based on this, overall conclusions are drawn on various levels and the contributions are elaborated. Finally, an outlook on future research directions is given.

5.1 Summary

This work proposes a multistage perception approach for railway environments that aims at reducing accuracy and ambiguity problems of current train-borne localization systems. It solely bases on mobile lidar sensor measurements and thus relies no longer on costly track-side infrastructure. The basic idea is to detect rails as the pervasive key element, which are fundamental for the running of a railway vehicle. Since positioning ambiguities primarily result from nearby tracks and their connections with the ego track, those elements are further detected.

The corresponding method is based on both features and model knowledge, which highly reflect the structure of a common railway network and its characteristics. Occluding edges, significant deviations of the echo, and a template matching allow the detection of rail profiles. Rail and track sections can be determined with the help of spatially clustered detections and their associations. From their geometrical and chronological arrangement, information on the nearby environment in general (railway network topology) and in particular branching maneuvers of the railway vehicle on single turnouts (position, direction, and side taken thereon) can be derived.

The approach achieved convincing qualitative and quantitative results. Its extensive evaluation on a demanding test ground covered varied topologies with neighboring tracks and several single turnouts, which enabled to test the approach on more than 200 branching situations at different velocities. Finally, it has been shown which characteristics of the proposed topology and branching events are required for an integration into a train-borne localization system and how their potential inaccuracy or incompleteness is considered.

5.2 Conclusions

Conclusions can be drawn on various levels, which include the suitability of the proposed setup, the extent of the detected objects, the achieved results, and their utilization within a self-localization system.
The results show that even a low cost lidar sensor, which is mounted on the railway vehicle, is capable of perceiving the relevant nearby environment and is thereby independent of illumination conditions. Furthermore, only distance and echo measurements within vertical cross-sections, as well as their temporal and spatial arrangement, are adequate for the detections, so that additional sensors with their individual error sources can be waived.
The approach utilizes ideas and methods proposed in related work. This allows the detection of rails as well as the ego track and neighboring tracks. However, in contrast to many other approaches, turnouts are explicitly considered and even the branching direction thereon is detected.
Furthermore, the coverage of the experimental evaluation exceeds most similar works. The achieved results are promising, even though a few weaknesses occur, such as the fact that occluded rails cannot be detected on level crossings, or that a few false positive turnout detections arise. Even so, the strengths of the proposed approach are demonstrated, for example by an accurate and reproducible determination of the tip of the blade, which has an absolute deviation of less than half a meter in more than 90 % of all cases, or the correctness of topology and branching events, which exceeds 98 %. Thus, these results allow

to adjust the estimated position of the railway vehicle and assess their probabilities in the case of ambiguity within a train-borne localization system.
Finally and with the help of [Lauer and Stein 2015; Pottberg 2016], it has been demonstrated that topology and branching events can be modeled stochastically and in this way integrated into such a self-localization system. While state of the art systems, which base on measurements of GNSS, an odometer or tachometer, and a digital map, suffer from position inaccuracy and missing track-selectivity, this joint approach can reduce or even eliminate their negative effects.

5.3 Future work

Based on the results of this work, it is now possible to further develop the research prototype for lidar based infrastructure detection to a system on product level and to integrate it into an on-board self-localization approach for railway vehicles. Different from the development of this work, which was designed as a pure research project, the systematic product development could achieve the required safety level for operation in safety relevant environments. The application of such a system would finally allow to remove track-side infrastructure for positioning of railway vehicles and increase the capacity of railway lines. The first steps into that direction have already been developed in the GaLoROI project [Manz et al. 2015].
Moreover, the further development of self-localization approaches for railway scenarios requires an extensive empirical evaluation in order to prove the reliability and safety of these systems. The amount of required data considerably exceeds all experiments that have been done by now. Thus, the systematic recording of sensor data for self-localization in a large scale experiment becomes an important issue in the future. First steps into this direction have been started in a collaboration project between Deutsche Bahn and the Technische Universität Darmstadt which is strongly supported by the railway research group at the Institute of Measurement and Control Systems of the Karlsruhe Institute of Technology [Winter et al. 2017].

Finally, although this work has already shown how rails and turnouts are identified within lidar sensor measurements and how their detections can support a self-localization approach, there are further objects next to the track that might be detected with the mobile laser scanning system, e.g., level crossings, tunnels, bridges, and platforms. While those objects do not belong to the track itself, they might serve as additional reference points and thus, provide valuable information for a train-borne localization system.

References

Adams, R. and L. Bischof (1994). Seeded region growing. In: *IEEE Trans. Pattern Anal. Mach. Intell.* 16 (6), pp. 641–647.

Allotta, B., P. D'Adamio, M. Malvezzi, L. Pugi, A. Ridolfi, A. Rindi, and G. Vettori (2014). An innovative localisation algorithm for railway vehicles. In: *Vehicle System Dynamics* 52 (11), pp. 1443–1469.

Arastounia, M. (2015). Automated Recognition of Railroad Infrastructure in Rural Areas from LIDAR Data. In: *Remote Sensing* 7 (11), pp. 14916–14938.

Arastounia, M. (2017). An Enhanced Algorithm for Concurrent Recognition of Rail Tracks and Power Cables from Terrestrial and Airborne LiDAR Point Clouds. In: *Infrastructures* 2 (2), no. 8.

Arastounia, M. and S. Oude Elberink (2016). Application of Template Matching for Improving Classification of Urban Railroad Point Clouds. In: *Sensors* 16 (12), no. 2112.

Bacci, G., E. Falletti, C. Fernández-Prades, M. Luise, D. Margaria, and F. Zanier (2012). Satellite-Based Navigation Systems. In: *Satellite and Terrestrial Radio Positioning Techniques*. Ed. by D. Dardari, E. Falletti, and M. Luise, pp. 25–74.

Beger, R., C. Gedrange, R. Hecht, and M. Neubert (2011). Data fusion of extremely high resolution aerial imagery and LiDAR data for automated railroad centre line reconstruction. In: *ISPRS J. Photogrammetry and Remote Sensing* 66 (6, Suppl.), pp. S40–S51.

Blug, A., C. Baulig, H. Wölfelschneider, and H. Höfler (2004). Fast fiber coupled clearance profile scanner using real time 3D data processing with automatic rail detection. In: *Proc. IEEE Intell. Veh. Symp.* pp. 658–663.

BOStrab. *Straßenbahn-Bau- und Betriebsordnung vom 11.12.1987 (BGBl. I S. 2648), die zuletzt durch Artikel 1 der Verordnung vom 16.12.2016 (BGBl. I S. 2938) geändert worden ist.*

Broquetas, A., A. Comerón, A. Gelonch, J. M. Fuertes, J. A. Castro, D. Felip, M. A. López, and J. A. Pulido (2012). Track Detection in Railway Sidings Based on MEMS Gyroscope Sensors. In: *Sensors* 12 (12), pp. 16228–16249.

Corsino Espino, J. and B. Stanciulescu (2013). Turnout detection and classification using a modified HOG and template matching. In: *Proc. Int. IEEE Conf. Intell. Transp. Syst.* pp. 2045–2050.

Daiß, K. (2014). Schätzung der Gleisgeometrie aus 3D-Laserscandaten. Bachelor thesis. Karlsruhe Institute of Technology.

Daoust, T., F. Pomerleau, and T. D. Barfoot (2016). Light at the End of the Tunnel: High-Speed LiDAR-Based Train Localization in Challenging Underground Environments. In: *Proc. Conf. Comput. and Robot Vision*, pp. 93–100.

EBO. *Eisenbahn-Bau- und Betriebsordnung vom 08.05.1967 (BGBl. 1967 II S. 1563), die durch Artikel 174 des Gesetzes vom 29.03.2017 (BGBl. I S. 626) geändert worden ist.*

EN 13232-2 (2003). Railway applications - Track - Switches and Crossings - Requirements for geometric design. Standard.

EN 13232-3 (2003). Railway applications - Track - Switches and Crossings - Requirements for wheel/rail interaction. Standard.

EN 13232-5 (2005). Railway applications - Track - Switches and Crossings - Switches. Standard.

EN 13232-6 (2005). Railway applications - Track - Switches and Crossings - Fixed common and obtuse crossings. Standard.

EN 13674-1 (2011). Railway applications - Track - Rail - Vignole railway rails 46 kg/m and above. Standard.

EN 13674-4 (2006). Railway applications - Track - Rail - Vignole railway rails from 27 kg/m to, but excluding 46 kg/m. Standard.

EN 13848-1 (2003). Railway applications - Track - Track geometry quality - Characterisation of track geometry. Standard.

EN 13848-2 (2006). Railway applications - Track - Track geometry quality - Measuring systems - Track recording vehicles. Standard.

EN 14811 (2006). Railway applications - Track - Special purpose rail - Grooved and associated construction. Standard.

EN 50121 (2015). Railway applications - Electromagnetic compatibility. Standards series.

EN 50126 (1999). Railway applications - The specification and demonstration of reliability, availability, maintainability and safety (RAMS). Standard.

ESBO. *Eisenbahn-Bau- und Betriebsordnung für Schmalspurbahnen vom 25.02. 1972 (BGBl. I S. 269), die zuletzt durch Artikel 519 der Verordnung vom 31.08.2015 (BGBl. I S. 1474) geändert worden ist.*

European Commission (2014). EU transport Scoreboard: Electrified railway lines. `https://ec.europa.eu/transport/facts-fundings/scoreboard/compare/energy-union-innovation/share-electrified-railway_en`. Accessed: 26.06.2017.

Federal Railroad Administration (2017). Office of Safety Analysis: 3.02 - Accident Trends - Graphs & Charts - Train Accidents by Primary Cause. `http://safetydata.fra.dot.gov/OfficeofSafety/publicsite/graphs.aspx`. Accessed: 26.06.2017.

Fiedler, J. (1999). *Bahnwesen: Planung, Bau und Betrieb von Eisenbahnen, S-, U-, Stadt- und Straßenbahnen*. 4th ed.

Fischler, M. A. and R. C. Bolles (1981). Random Sample Consensus: A Paradigm for Model Fitting with Applications to Image Analysis and Automated Cartography. In: *Commun. ACM* 24 (6), pp. 381–395.

Geistler, A. and F. Böhringer (2004). Detection And Classification Of Turnouts Using Eddy Current Sensors. In: *Comput. in Railways IX*. Ed. by J. Allan, C. A. Brebbia, R. J. Hill, G. Sciutto, and S. Sone. Vol. 74. WIT Trans. The Built Environment, pp. 467–476.

Hackel, T. (2013). Erkennung von Weichenbestandteilen aus 3D-Laserscandaten. Diploma thesis. Karlsruhe Institute of Technology.

Hackel, T., D. Stein, I. Maindorfer, M. Lauer, and A. Reiterer (2015). Track detection in 3D laser scanning data of railway infrastructure. In: *Proc. IEEE Int. Instrum. and Meas. Technol. Conf.* pp. 693–698.

Heirich, O., A. Steingass, A. Lehner, and T. Strang (2013a). Velocity and location information from onboard vibration measurements of rail vehicles. In: *Proc. Int. Conf. Inform. Fusion*, pp. 1835–1840.

Heirich, O., A. Lehner, P. Robertson, and T. Strang (2011). Measurement and analysis of train motion and railway track characteristics with inertial sensors. In: *Proc. Int. IEEE Conf. Intell. Transp. Syst.* pp. 1995–2000.

Heirich, O., P. Robertson, A. Cardalda García, and T. Strang (2012). Bayesian Train Localization Method Extended By 3D Geometric Railway Track Observations From Inertial Sensors. In: *Proc. Int. Conf. Inform. Fusion*, pp. 416–423.

Heirich, O., P. Robertson, and T. Strang (2013b). RailSLAM - Localization of Rail Vehicles and Mapping of Geometric Railway Tracks. In: *Proc. IEEE Int. Conf. Robot. and Automat.* pp. 5192–5199.

Heirich, O. and B. Siebler (2015). Train-side passive magnetic measurements. In: *Proc. IEEE Int. Instrum. and Meas. Technol. Conf.* pp. 687–692.

Hensel, S., C. Hasberg, and C. Stiller (2011). Probabilistic Rail Vehicle Localization With Eddy Current Sensors in Topological Maps. In: *IEEE Trans. Intell. Transp. Syst.* 12 (4), pp. 1525–1536.

Johannes, L. and E. Almeida (2014). Zugortung mittels Laserscanner (Train positioning with laser scanners). In: *ZEVrail*, pp. 450–457.

Jung, J., L. Chen, G. Sohn, C. Luo, and J.-U. Won (2016). Multi-Range Conditional Random Field for Classifying Railway Electrification System Objects Using Mobile Laser Scanning Data. In: *Remote Sensing* 8 (12), no. 1008.

Jwa, Y. and G. Sohn (2015). Kalman Filter Based Railway Tracking from Mobile LiDAR Data. In: *ISPRS Ann. Photogrammetry, Remote Sensing and Spatial Inform. Sci.* II/3-W5, pp. 159–164.

Kaleli, F. and Y. S. Akgul (2009). Vision-based railroad track extraction using dynamic programming. In: *Proc. Int. IEEE Conf. Intell. Transp. Syst.* pp. 42–47.

Kang, Z., L. Zhang, L. Tuo, B. Wang, and J. Chen (2014). Continuous Extraction of Subway Tunnel Cross Sections Based on Terrestrial Point Clouds. In: *Remote Sensing* 6 (1), pp. 857–879.

Kremer, J. and A. Grimm (2012). The RailMapper - A dedicated mobile lidar mapping system for railway networks. In: *ISPRS Int. Archives Photogrammetry, Remote Sensing and Spatial Inform. Sci.* XXXIX-B5, pp. 477–482.

Kutterer, H. (2010). Mobile Mapping. In: *Airborne and Terrestrial Laser Scanning*. Ed. by G. Vosselman and H. G. Maas, pp. 293–311.

Lauer, M. and D. Stein (2013). Algorithms and concepts for an onboard train localization system for safety-relevant services. In: *Proc. IEEE Int. Conf. Intell. Rail Transp.* pp. 65–70.

Lauer, M. and D. Stein (2015). A Train Localization Algorithm for Train Protection Systems of the Future. In: *IEEE Trans. Intell. Transp. Syst.* 16 (2), pp. 970–979.

Leslar, M., G. Perry, and K. McNease (2010). Using mobile lidar to survey a railway line for asset inventory. In: *Proc. ASPRS Ann. Conf.*

Lüdicke, D., F. Eßer, T. Marchand, and T. Dellmann (2014). Modular Multi-domain Co-simulation for Rail Vehicle Testing with ETCS Scenario Control (WIP). In: *Proc. Summer Simulation Multi-Conf.* pp. 7:1–7:9.

Maire, F. (2007). Vision based anti-collision system for rail track maintenance vehicles. In: *Proc. IEEE Conf. Advanced Video and Signal Based Surveillance*, pp. 170–175.

Manz, H., E. Schnieder, D. Stein, M. Spindler, M. Lauer, C. Seedorff, A. Baudis, U. Becker, J. Beugin, T. K. Nguyen, and J. Marais (2015). GaLoROI: Satellite based localization in railways. In: *Proc. Int. Congr. Advanced Railway Eng.* pp. 322–329.

Marais, J., J. Beugin, and M. Berbineau (2017). A Survey of GNSS-Based Research and Developments for the European Railway Signaling. In: *IEEE Trans. Intell. Transp. Syst.* 18 (10), pp. 2602–2618.

Matthews, V. (2007). *Bahnbau.* 7th ed.

Mesch, F., F. Puente León, and T. Engelberg (2000). Train-based location by detecting rail switches. In: *Comput. in Railways VII*. Ed. by J. Allan, R. J. Hill, C. A. Brebbia, G. Sciutto, and S. Sone. Vol. 50. WIT Trans. The Built Environment, pp. 1251–1260.

Mikrut, S., P. Kohut, K. Pyka, R. Tokarczyk, T. Barszcz, and T. Uhl (2016). Mobile Laser Scanning Systems for Measuring the Clearance Gauge of Railways: State of Play, Testing and Outlook. In: *Sensors* 16 (5), no. 683.

Mirabadi, A., N. Mort, and F. Schmid (1996). Application of sensor fusion to railway systems. In: *IEEE/SICE/RSJ Int. Conf. Multisensor Fusion and Integration for Intell. Syst.* pp. 185–192.

Mohamad, M., K. Kusevic, P. Mrstik, and M. Greenspan (2013). Automatic Rail Extraction in Terrestrial and Airborne LiDAR Data. In: *Proc. Int. Conf. 3D Vision*, pp. 303–309.

Nash, A., D. Huerlimann, J. Schuette, and V. P. Krauss (2004). RailML - A Standard Data Interface For Railroad Applications. In: *Comput. in Railways IX*. Ed. by J. Allan, C. A. Brebbia, R. J. Hill, G. Sciutto, and S. Sone. Vol. 74. WIT Trans. The Built Environment, pp. 233–240.

Neubert, M., R. Hecht, C. Gedrange, M. Trommler, H. Herold, T. Krüger, and F. Brimmer (2008). Extraction of railroad objects from very high resolution helicopter-borne LiDAR and ortho-image data. In: *ISPRS Int. Archives Photogrammetry, Remote Sensing and Spatial Inform. Sci.* XXXVIII-4/C1, pp. 25–30.

Oude Elberink, S. and K. Khoshelham (2015). Automatic Extraction of Railroad Centerlines from Mobile Laser Scanning Data. In: *Remote Sensing* 7 (5), pp. 5565–5583.

Oude Elberink, S., K. Khoshelham, M. Arastounia, and D. Díaz Benito (2013). Rail track detection and modelling in mobile laser scanner data. In: *ISPRS*

Ann. Photogrammetry, Remote Sensing and Spatial Inform. Sci. II/5-W2, pp. 223–228.

Pachl, J. (2016). *Systemtechnik des Schienenverkehrs: Bahnbetrieb planen, steuern und sichern.* 8th ed.

Pepperl+Fuchs GmbH (2016a). Manual: 2-D Laser Scanner OMD10M-R2000-B23, OMD30M-R2000-B23, OMD12M-R2000-B23.

Pepperl+Fuchs GmbH (2016b). Operation Instructions: OMDxxx-R2000 Ethernet communication protocol - Protocol version 1.02.

Plan, O. (2003). GIS-gestützte Verfolgung von Lokomotiven im Werkbahnverkehr. PhD thesis. Universität der Bundeswehr München.

Pólya, Z. M. (2013). Erkennung von Gleisen und deren Bestandteilen in Laserscandaten. Internship report. Karlsruhe Institute of Technology.

Ponciano, J.-J., C. Prudhomme, B. Tietz, and F. Boochs (2015). Detection and isolation of switches in point clouds of the German railway network. In: *Proc. Int. Conf. Signal-Image Technol. Internet-Based Syst.* pp. 96–102.

Pottberg, S. (2016). Probabilistic Modeling of Track-Events for an Onboard Train Localization System. Master thesis. Karlsruhe Institute of Technology.

Puente, I., H. González-Jorge, J. Martínez-Sánchez, and P. Arias (2013). Review of mobile mapping and surveying technologies. In: *Meas.* 46 (7), pp. 2127–2145.

Qi, Z., Y. Tian, and Y. Shi (2013). Efficient railway tracks detection and turnouts recognition method using HOG features. In: *Neural Computing and Applications* 23 (1), pp. 245–254.

Rahmig, C. and L. Johannes (2013). Verfahren zur automatischen Detektion von charakteristischen Elementen, insbesondere eines Bahnübergangs, und Einrichtung dafür. Publication of application (Offenlegungsschrift) DE 1020131 04088 A1.

Rahmig, C., L. Johannes, and K. Lüddecke (2013). Detecting Track Events with a Laser Scanner for using within a Modified Multi-Hypothesis Based Map-Matching Algorithm for Train Positioning. In: *Proc. European Navigation Conf.*

Rahmig, C. and A. Kluge (2013). Digital maps for railway applications based on OpenStreetMap data. In: *Proc. Int. IEEE Conf. Intell. Transp. Syst.* pp. 1322–1327.

Reiterer, A., H. Höfler, H. Wölfelschneider, C. Baulig, I. Maindorfer, N. Dimopoulos, S. Schwarzer, and M. Dambacher (2014). Railway Measurement Techniques: Opportunities and Challenges. In: *Proc. Int. Conf. Railway Technol.: Research, Development and Maintenance*, no. 86.

Riehm, B. J. (2013). Lidar-basierte Landmarkenerkennung zur verbesserten Selbstlokalisierung von Schienenfahrzeugen. Bachelor thesis. Karlsruhe Institute of Technology.

Röver, S., M. Schroeder, and E. Schnieder (1998). Satellite based locating in guided traffic. In: *Comput. in Railways VI.* Ed. by B. Mellitt, R. J. Hill, J. Allan, G. Sciutto, and C. A. Brebbia. Vol. 37. WIT Trans. The Built Environment, pp. 659–668.

Saab, S. S. (2000a). A map matching approach for train positioning. Part I: Development and analysis. In: *IEEE Trans. Veh. Technol.* 49 (2), pp. 467–475.

Saab, S. S. (2000b). A map matching approach for train positioning. Part II: Application and experimentation. In: *IEEE Trans. Veh. Technol.* 49 (2), pp. 476–484.

Siemens AG (2012). Reliable rail traffic - from the very first day: Test- and Validationcenter Wegberg-Wildenrath.

Siemens AG (2016). ICE 4 (BR 412): high-speed trains.

Soni, A., S. Robson, and B. Gleeson (2014). Extracting Rail Track Geometry from Static Terrestrial Laser Scans for Monitoring Purposes. In: *ISPRS Int. Archives Photogrammetry, Remote Sensing and Spatial Inform. Sci.* XL-5, pp. 553–557.

Spindler, M., D. Stein, and M. Lauer (2016a). Difference-Inductance-Sensor zur berührungslosen Geschwindigkeitsbestimmung von Zügen. In: *XXX. Messtechnisches Symp.* Ed. by S. Zimmermann, pp. 193–200.

Spindler, M., D. Stein, and M. Lauer (2016b). Low Power and Low Cost Sensor for Train Velocity Estimation. In: *Proc. IEEE Int. Conf. Intell. Rail Transp.* pp. 259–264.

Stadlmann, B., S. Mairhofer, and G. Hanis (2010). Field Experience with GPS based Train Control System. In: *Proc. European Navigation Conf. on Global Navigation Satellite Syst.*

Stein, D., M. Lauer, and M. Spindler (2014a). An analysis of different sensors for turnout detection for train-borne localization systems. In: *Comput. in Railways XIV: Railway Eng. Design and Optimization.* Ed. by C. A. Brebbia, N. Tomii, P. Tzieropoulos, and J. M. Mera. Vol. 135. WIT Trans. The Built Environment, pp. 827–838.

Stein, D., M. Spindler, J. Kuper, and M. Lauer (2016a). Rail detection using lidar sensors. In: *Int. J. Sustainable Development and Planning* 11 (1), pp. 65–78.

Stein, D., M. Spindler, and M. Lauer (2014b). Lidar sensors for detecting railway infrastructure and their usage in train-borne localization systems. In: Poster: MoLaS Technol. Workshop.

Stein, D., M. Spindler, and M. Lauer (2016b). Model-based rail detection in mobile laser scanning data. In: *Proc. IEEE Intell. Veh. Symp.* pp. 654–661.

Winter, H., V. Willert, J. Adamy, M. Leining, M. Spindler, M. Lauer, D. Stein, O. Heirich, J. Groos, A. Geffert, U. Becker, and M. Breuer (2017). Localization Reference Train - Sichere Ortung für den Schienenverkehr. In: *Proc. Scientific Railway Signalling Symp.* in press.

Wohlfeil, J. (2011). Vision based rail track and switch recognition for self-localization of trains in a rail network. In: *Proc. IEEE Intell. Veh. Symp.* pp. 1025–1030.

Wölfelschneider, H. (2009). Physikalische Prinzipien der Laserscantechnologie. In: *Terrestrisches Laserscanning (TLS 2009). Yes, we Scan!* Ed. by F. Godhoff and R. Staiger, pp. 13–30.

Yang, B. and L. Fang (2014). Automated Extraction of 3-D Railway Tracks from Mobile Laser Scanning Point Clouds. In: *IEEE J. Sel. Topics in Appl. Earth Observ. and Remote Sens.* 7 (12), pp. 4750–4761.

Zwemer, M. H., D. W. J. M. van de Wouw, E. Jaspers, S. Zinger, and P. H. N. de With (2015). A vision-based approach for tramway rail extraction. In: *Proc. SPIE* 9407, no. 94070R.

Publications by the author

Hackel, T., D. Stein, I. Maindorfer, M. Lauer, and A. Reiterer (2015). Track detection in 3D laser scanning data of railway infrastructure. In: *Proc. IEEE Int. Instrum. and Meas. Technol. Conf.* pp. 693–698.

Lauer, M. and D. Stein (2013). Algorithms and concepts for an onboard train localization system for safety-relevant services. In: *Proc. IEEE Int. Conf. Intell. Rail Transp.* pp. 65–70.

Lauer, M. and D. Stein (2015). A Train Localization Algorithm for Train Protection Systems of the Future. In: *IEEE Trans. Intell. Transp. Syst.* 16 (2), pp. 970–979.

Manz, H., E. Schnieder, D. Stein, M. Spindler, M. Lauer, C. Seedorff, A. Baudis, U. Becker, J. Beugin, T. K. Nguyen, and J. Marais (2015). GaLoROI: Satellite based localization in railways. In: *Proc. Int. Congr. Advanced Railway Eng.* pp. 322–329.

Spindler, M., D. Stein, and M. Lauer (2016a). Difference-Inductance-Sensor zur berührungslosen Geschwindigkeitsbestimmung von Zügen. In: *XXX. Messtechnisches Symp.* Ed. by S. Zimmermann, pp. 193–200.

Spindler, M., D. Stein, and M. Lauer (2016b). Low Power and Low Cost Sensor for Train Velocity Estimation. In: *Proc. IEEE Int. Conf. Intell. Rail Transp.* pp. 259–264.

Stein, D., M. Lauer, and M. Spindler (2014a). An analysis of different sensors for turnout detection for train-borne localization systems. In: *Comput. in Railways XIV: Railway Eng. Design and Optimization.* Ed. by C. A. Brebbia, N. Tomii, P. Tzieropoulos, and J. M. Mera. Vol. 135. WIT Trans. The Built Environment, pp. 827–838.

Stein, D., M. Spindler, J. Kuper, and M. Lauer (2016a). Rail detection using lidar sensors. In: *Int. J. Sustainable Development and Planning* 11 (1), pp. 65–78.

Stein, D., M. Spindler, and M. Lauer (2014b). Lidar sensors for detecting railway infrastructure and their usage in train-borne localization systems. In: Poster: MoLaS Technol. Workshop.

Stein, D., M. Spindler, and M. Lauer (2016b). Model-based rail detection in mobile laser scanning data. In: *Proc. IEEE Intell. Veh. Symp.* pp. 654–661.

Winter, H., V. Willert, J. Adamy, M. Leining, M. Spindler, M. Lauer, D. Stein, O. Heirich, J. Groos, A. Geffert, U. Becker, and M. Breuer (2017). Localization Reference Train - Sichere Ortung für den Schienenverkehr. In: *Proc. Scientific Railway Signalling Symp.* in press.

Supervised theses

Daiß, K. (2014). Schätzung der Gleisgeometrie aus 3D-Laserscandaten. Bachelor thesis. Karlsruhe Institute of Technology.

Hackel, T. (2013). Erkennung von Weichenbestandteilen aus 3D-Laserscandaten. Diploma thesis. Karlsruhe Institute of Technology.

Pólya, Z. M. (2013). Erkennung von Gleisen und deren Bestandteilen in Laserscandaten. Internship report. Karlsruhe Institute of Technology.

Pottberg, S. (2016). Probabilistic Modeling of Track-Events for an Onboard Train Localization System. Master thesis. Karlsruhe Institute of Technology.

Riehm, B. J. (2013). Lidar-basierte Landmarkenerkennung zur verbesserten Selbstlokalisierung von Schienenfahrzeugen. Bachelor thesis. Karlsruhe Institute of Technology.

Schriftenreihe
Institut für Mess- und Regelungstechnik
Karlsruher Institut für Technologie
(1613-4214)

Band 001 Hans, Annegret
Entwicklung eines Inline-Viskosimeters auf Basis eines magnetisch-induktiven Durchflussmessers. 2004
ISBN 3-937300-02-3

Band 002 Heizmann, Michael
Auswertung von forensischen Riefenspuren mittels automatischer Sichtprüfung. 2004
ISBN 3-937300-05-8

Band 003 Herbst, Jürgen
Zerstörungsfreie Prüfung von Abwasserkanälen mit Klopfschall. 2004
ISBN 3-937300-23-6

Band 004 Kammel, Sören
Deflektometrische Untersuchung spiegelnd reflektierender Freiformflächen. 2005
ISBN 3-937300-28-7

Band 005 Geistler, Alexander
Bordautonome Ortung von Schienenfahrzeugen mit Wirbelstrom-Sensoren. 2007
ISBN 978-3-86644-123-1

Band 006 Horn, Jan
Zweidimensionale Geschwindigkeitsmessung texturierter Oberflächen mit flächenhaften bildgebenden Sensoren. 2007
ISBN 978-3-86644-076-0

Die Bände sind unter www.ksp.kit.edu als PDF frei verfügbar oder als Druckausgabe bestellbar.

Band 007 Hoffmann, Christian
Fahrzeugdetektion durch Fusion monoskopischer Videomerkmale. 2007
ISBN 978-3-86644-139-2

Band 008 Dang, Thao
Kontinuierliche Selbstkalibrierung von Stereokameras. 2007
ISBN 978-3-86644-164-4

Band 009 Kapp, Andreas
Ein Beitrag zur Verbesserung und Erweiterung der Lidar-Signalverarbeitung für Fahrzeuge. 2007
ISBN 978-3-86644-174-3

Band 010 Horbach, Jan
Verfahren zur optischen 3D-Vermessung spiegelnder Oberflächen. 2008
ISBN 978-3-86644-202-3

Band 011 Böhringer, Frank
Gleisselektive Ortung von Schienenfahrzeugen mit bordautonomer Sensorik. 2008
ISBN 978-3-86644-196-5

Band 012 Xin, Binjian
Auswertung und Charakterisierung dreidimensionaler Messdaten technischer Oberflächen mit Riefentexturen. 2009
ISBN 978-3-86644-326-6

Band 013 Cech, Markus
Fahrspurschätzung aus monokularen Bildfolgen für innerstädtische Fahrerassistenzanwendungen. 2009
ISBN 978-3-86644-351-8

Band 014 Speck, Christoph
Automatisierte Auswertung forensischer Spuren auf Patronenhülsen. 2009
ISBN 978-3-86644-365-5

Band 015 Bachmann, Alexander
Dichte Objektsegmentierung in Stereobildfolgen. 2010
ISBN 978-3-86644-541-3

Band 016 Duchow, Christian
Videobasierte Wahrnehmung markierter Kreuzungen mit lokalem Markierungstest und Bayes'scher Modellierung. 2011
ISBN 978-3-86644-630-4

Band 017 Pink, Oliver
Bildbasierte Selbstlokalisierung von Straßenfahrzeugen. 2011
ISBN 978-3-86644-708-0

Band 018 Hensel, Stefan
Wirbelstromsensorbasierte Lokalisierung von Schienenfahrzeugen in topologischen Karten. 2011
ISBN 978-3-86644-749-3

Band 019 Carsten Hasberg
Simultane Lokalisierung und Kartierung spurgeführter Systeme. 2012
ISBN 978-3-86644-831-5

Band 020 Pitzer, Benjamin
Automatic Reconstruction of Textured 3D Models. 2012
ISBN 978-3-86644-805-6

Band 021 Roser, Martin
Modellbasierte und positionsgenaue Erkennung von Regentropfen in Bildfolgen zur Verbesserung von videobasierten Fahrerassistenzfunktionen. 2012
ISBN 978-3-86644-926-8

Band 022 Loose, Heidi
Dreidimensionale Straßenmodelle für Fahrerassistenzsysteme auf Landstraßen. 2013
ISBN 978-3-86644-942-8

Band 023 Rapp, Holger
Reconstruction of Specular Reflective Surfaces using Auto-Calibrating Deflectometry. 2013
ISBN 978-3-86644-966-4

Band 024 Moosmann, Frank
Interlacing Self-Localization, Moving Object Tracking and Mapping for 3D Range Sensors. 2013
ISBN 978-3-86644-977-0

Band 025 Geiger, Andreas
Probabilistic Models for 3D Urban Scene Understanding from Movable Platforms. 2013
ISBN 978-3-7315-0081-0

Band 026 Hörter, Marko
Entwicklung und vergleichende Bewertung einer bildbasierten Markierungslichtsteuerung für Kraftfahrzeuge. 2013
ISBN 978-3-7315-0091-9

Band 027 Kitt, Bernd
Effiziente Schätzung dichter Bewegungsvektorfelder unter Berücksichtigung der Epipolargeometrie zwischen unterschiedlichen Ansichten einer Szene. 2013
ISBN 978-3-7315-0105-3

Band 028 Lategahn, Henning
Mapping and Localization in Urban Environments Using Cameras. 2013
ISBN 978-3-7315-0135-0

Band 029 Tischler, Karin
Informationsfusion für die kooperative Umfeldwahrnehmung vernetzter Fahrzeuge. 2014
ISBN 978-3-7315-0166-4

Band 030 Schmidt, Christian
Fahrstrategien zur Unfallvermeidung im Straßenverkehr für Einzel- und Mehrobjektszenarien. 2014
ISBN 978-3-7315-0198-5

Band 031 Firl, Jonas
Probabilistic Maneuver Recognition in Traffic Scenarios. 2014
ISBN 978-3-7315-0287-6

Band 032 Schönbein, Miriam
Omnidirectional Stereo Vision for Autonomous Vehicles. 2015
ISBN 978-3-7315-0357-6

Band 033 Nicht erschienen

Band 034 Liebner, Martin
Fahrerabsichtserkennung und Risikobewertung für warnende Fahrerassistenzsysteme. 2016
ISBN 978-3-7315-0508-2

Band 035 Ziegler, Julius
Optimale Trajektorienplanung für Automobile. 2017
ISBN 978-3-7315-0553-2

Band 036 Harms, Hannes
Genauigkeitsuntersuchung von binokularen Normalenvektoren für die Umfeldwahrnehmung. 2017
ISBN 978-3-7315-0628-7

Band 037 Ruhhammer, Christian
Inferenz von Kreuzungsinformationen aus Flottendaten. 2017
ISBN 978-3-7315-0721-5

Band 038 Stein, Denis
Mobile laser scanning based determination of railway network topology and branching direction on turnouts. 2018
ISBN 978-3-7315-0743-7